Air, Water, Earth, Fire

Angelo Peccerillo

Air, Water, Earth, Fire

How the System Earth Works

 Springer

Angelo Peccerillo
Firenze, Italy

ISBN 978-3-030-78015-9   ISBN 978-3-030-78013-5   (eBook)
https://doi.org/10.1007/978-3-030-78013-5

This Springer imprint is published by the registered company Springer Nature Switzerland AG
The registered company address is: Gewerbestrasse 11, 6330 Cham, Switzerland

Preface

From time to time, it happens that scientific discoveries go beyond the restricted circle of discipline specialists, being widely disseminated through the mass media. The news of science most frequently reported by the press and television generally deal with medicine, biology, and physics, particularly astronomy. By contrast, discoveries in the Earth Sciences rarely attract media attention, although they involve natural phenomena that directly impact the environment.

Recently, a team of British and Italian researchers demonstrated a temporal relationship between some crucial changes of life on Earth and intense modifications in the strength of the geomagnetic field. Geomagnetism provides protection to living organisms against the solar wind. The occurrence of a strong magnetic field enveloping our planet has allowed the development of the first microorganisms and might also have facilitated the so-called *Cambrian explosion*, i.e. the rapid diversification of multicellular animal life that occurred around the beginning of the Cambrian period at about 540 million years ago. In short, without the presence of an intense magnetic field, life on Earth would not exist.

One might expect that such important news would find adequate space in the public media; however, this was not the case, at least in Italy. As a result, most people ignore the paramount role of the geomagnetic field for the birth, evolution and survival of all organisms living on Earth, including the human species.

The same fate has befallen many other geological topics, including the long-running dispute about the increase of CO_2 in the Earth's atmosphere over the last century, a geological process of primary significance that, although much considered in the media, is rarely framed in its proper scientific context, that is, within the framework of the global geochemical cycle of carbon.

These and many other considerations have led me to conclude that the spread of scientific education does not value much the geosciences, a group of disciplines that investigate many crucial topics such as the chemical-physical characteristics, dynamics, and evolution of the Earth, and their role in constructing an external environment suitable for complex life, a unique case in the universe—as far as we presently know. Typically, geological phenomena are worthy of media attention only during natural disasters such as earthquakes, tsunamis, and volcanic eruptions.

In contrast with the dominant trend, I decided to write this book about the basics of geology, illustrating without pretension of completeness and originality some of the main geological processes that have shaped our planet, making it friendly for complex life. These notions are the standard cultural background of geologists, but I have reason to believe they are little known, if not wholly unknown, across the public at large. Therefore, this book aims at giving some basic scientific information on how our planet works, something that should belong to the cultural background of any conscious individual, in the author's opinion. Geologists will not find much interesting in the book, except perhaps the attempt to frame the most disparate of geological processes in a holistic context, and the particular perspective with which the various topics are addressed, sensibly oriented towards geochemistry.

Giuseppe Tomasi di Lampedusa (1896–1957), a highly regarded novelist from Sicily, stated:

> *Writing his memories and experiences at a certain age should be a duty imposed by the State to everyone. The material that would accumulate after three or four generations would be invaluable: many psychological and historical problems that beset humanity would be solved. There are no memories, although written by common people, which do not contain first-order values.*

If this is true for the personal experiences of ordinary people, the same should apply, and perhaps even more so, to scientists. Therefore, the great Sicilian novelist can be paraphrased by saying that if every scientist decides to write his experiences to transmit discoveries and illustrate what he understood about his science, then the spread of scientific culture would receive a decisive impetus.

This book consists of nine chapters and an Epilogue. The first three chapters provide the foundation by presenting the structure of the Earth, and by describing the sedimentary and magmatic processes that respectively operate at the surface and inside our planet. Chapters 4 and 5 focus on the physics of the Earth system, in particular on geomagnetism and seismicity. Chapters 6–8 synthesise the information from the previous chapters to present the plate tectonics theory and the global geochemical cycles, exploring their impact on the external environment and life. Chapter 9 reviews the history of Earth from its formation in the solar nebula to the present time. The Epilogue contains a few informal philosophical-scientific considerations about the unique nature of our planet and discusses how knowledge and thinking of the geological past can lead us to make sound choices in the future. A geological time scale aimed at guiding the reader through the Earth's deep time is reported at the end of the book.

A complete reading of the book, from beginning to end, provides a full picture of how our planet works. However, the various chapters are organised to be read in isolation and not necessarily in sequence. This choice requires repetition, which makes the text redundant at times. Moreover, since *"life is short and great is the prolixity of the world"* (Josè Saramago), I thought it appropriate to add a summary at the end of each chapter. A preliminary reading of the summaries may help get an overview of the book, a first step in reading individual chapters. Detailed information boxes are added at the end of all chapters to go somewhat deeper into some key topics. These are not indispensable in understanding the essence of geological processes, but are intended to satisfy the curiosity of those who want to know more about specific subjects. Although the presentation is general in nature, various references to specialist publications are reported as footnotes to guide those who wish to investigate particular concepts in detail. Suggestions for further reading are reported at the end of the book.

A necessary price to pay for using a popular scientific approach and language is to lower the rigour of the discussion, leave out many important details, and insert elements that may appear purely ornamental. The following pages do not escape this rule, although I have done my best to make the tribute to simplification minimal, at least from the standpoint of scientific accuracy.

I express my deepest thanks to the colleagues and friends Russell S. Harmon and Carlo Bartolini for the critical reading of the manuscript and for suggestions and corrections.

I dedicate this work to my grandson, Alessandro Leonardo. It will be up to his generation to face the consequences on the environment of our choices and behaviours.

Firenze, Italy   Angelo Peccerillo

The original version of the book was revised: Copyright page text has been updated. The correction to the book is available at https://doi.org/10.1007/978-3-030-78013-5_11

Contents

About the Author

Angelo Peccerillo graduated in Geology cum laude, has been full professor of Petrology at the Universities of Messina, Cosenza and Perugia, where it has been teaching courses of Igneous petrology, Volcanology and Geochemistry (retired 2013). He has been editor or member of the editorial board of several international and national journals such as the European Journal of Mineralogy, Lithos, Journal of Volcanology and Geothermal Research, Open Mineralogy Journal, Journal of Virtual Explorer, and Journal of the Italian Geological Society. He has been topic editor, for Igneous and Metamorphic Petrology, of the Encyclopedia of Life Support System (EOLSS) printed by UNESCO. His scientific research has been focused on petrology, geochemistry and geodynamic significance of igneous processes and their role in the evolution of the Earth. For his activity he has been awarded the Feltrinelli Medal by the Accademia Nazionale dei Lincei, Rome, for the year 2006. He is author or co-author of some 250 scientific papers, mostly published on peer review international journals, several popular and didactic publications and scientific books, published with both national and international printers. He is Member of Academia Europaea and Honorary Member of the Geological Society of Italy.

1

The World Hidden Beneath Us—Structure and Composition of the Earth

Imagine there's no heaven
No hell below us
Above us only sky.
John Lennon - Imagine (1971)

1.1 Introduction

Elementa dum quattuor sunt: aer, aqua, tellus, ignis[1]: this is the aphorism with which scholars summarised the composition of the world, until a few centuries ago. Today, we know that the number of naturally occurring chemical elements is much higher; even so, this simplistic idea of ancient naturalists is still valid when applied to the Earth system as a whole.

The Earth can be viewed as consisting of four fundamental "ingredients" or spheres: the atmosphere (*aer*), hydrosphere (*aqua*), rocks and soil (*tellus*), and hot liquid materials such as magmas (*ignis*). However, the sphere of fire is not located between the Terrestrial Paradise and the Sky of the Moon, as envisioned by the Medieval cosmology, but is instead seated deep inside the Earth and gives continuous proof of its existence through volcanic activity.

While much is known about the atmosphere and hydrosphere (the so-called **fluid Earth**), the interior of the planet has been, and largely remains,

---

[1] Elements then are four: air, water, earth, fire.

© The Author(s), under exclusive license to Springer Nature
Switzerland AG 2021, corrected publication 2022
A. Peccerillo, *Air, Water, Earth, Fire,*
https://doi.org/10.1007/978-3-030-78013-5_1

unknown. Its inaccessibility has given rise to myths and phobias throughout the ages, from the Hades of Greek mythology and the Medieval Christian hell to the world of the Cimmerians and the mysterious realm of Agartha. For a long time, the dominant idea of a "Hollow Earth" thoroughly crossed by underground voids through which an internal fire rushed about in interconnected channels, was popular among ordinary folk, scientists and poets, as indicated in the writings of the Greek Homer, the Roman Pliny the Elder, Martianus Capella in the fifth century, Athanasius Kircher in the 17th and Jean-André Deluc in the eighteenth century, and into the early nineteenth century as postulated in Sir Humphrey Davy's model of volcanism.

Studies conducted in the last century got rid of these ideas by demonstrating that the body of the Earth is made up entirely of closely packed and compact rocks and liquids subjected to high pressure and temperature, progressively increasing with depth. The planet is stratified according to physical and chemical properties, especially density, and each layer or shell has its unique features. Such a structure is the outcome of a combination of geological processes, which have affected the Earth since its birth about 4.5 billion years ago, and are still going on today.

The various layers are not steady and isolated from each other, but rather move and continuously exchange matter and energy, making the Earth an active and continually evolving planet. Underground cavities and rivers do exist, but they only occur in certain rocks near the surface: a scientific revolution that has radically changed the view of the world hidden beneath us.

Knowledge of the interior of the Earth is a prerequisite for understanding geological processes. Since a large part of the Earth is formed of rocks (the so-called **solid Earth**), which, in turn, are composed of minerals, it is also necessary to have essential information on these crucial natural objects, which is summarised in Box 1.1.

The present structure of the Earth is the outcome of a countless number of events that have been going on for billions of years, from the aggregation of the planet about 4,600 million years ago to the present. Such a long history has been subdivided into various intervals on the basis of key geological events and their time of occurrence. The sequence of chronological units and subunits are grouped in the Geological Time Scale. A simplified version is reported at the end of the book and discussed in Chap. 9.

Fig. 1.1 **a** Xenoliths in volcanic rock at Torre Alfina, Province of Viterbo, Central Italy. Xenoliths are visible on the castle walls built with volcanic rocks quarried from a local eruptive centre. The xenoliths of Torre Alfina include a wide variety of rock types, such as peridotite, shales, schists, sandstones, limestone, and marls that represent samples torn out from the subsurface by rapidly ascending magmas; **b** Peridotite xenolith enclosed in a basalt

1.2 Xenoliths, Meteorites, Earthquakes: Witnesses of the Underground World

Most of the information on the structure and composition of the Earth's interior comes from studies of rocks and meteorites, and from the behaviour of the seismic waves that propagate through the planet during earthquakes. Direct observation of rocks is only possible for materials cropping out at the surface or residing at shallow depths where they can be accessed by mining, tunnelling, or drilling.[2] There is, however, a natural process that picks up rocks from the interior of the Earth and takes them to the surface, allowing us to obtain direct knowledge into deep regions that would otherwise be inaccessible. This "service", so to speak, is provided by a particular type of magma that originates at depths of 100–200 km and rises quickly to the eruption sites, tearing away and bringing to the surface fragments of rocks encountered during the ascent. These fragments are called **xenoliths** (from the Greek words ξένος λίθος, xénos lithos = foreign rock) and are easily distinguished from the host volcanic rock because of their contrasting colour and texture. Examples are reported in Fig. 1.1.

A particularly relevant class of xenoliths is called **peridotite**. This is a beautiful green crystalline rock (Fig. 1.1b), consisting of transparent green-coloured olivine crystals (or peridot), plus deep-green pyroxene and minor amounts of other minerals, mainly colourless plagioclase, black spinel

---

[2] The Kola Superdeep Borehole in the Kola Peninsula of Russia is the deepest yet hole made into the Earth's crust. It reached 12.2 km, which is only one third of the average thickness of the Earth's crust.

(MgAl$_2$O$_4$) or red garnet [Mg$_3$Al$_2$(SiO$_4$)$_3$], and sometimes diamond (see Box 1.1). These latter four minerals are stable, i.e. they can crystallise and remain unmodified, within rather narrow pressure ranges. Plagioclase is stable at pressures below one gigapascal[3] (P < 1 GPa); spinel occurs between 1 and 2 GPa; garnet and diamond are stable at higher pressures. Considering that pressure inside the Earth increases with depths by about 0.1 GPa (1 kbar) every 3.3–3.5 km due to the weight of the overlying rocks (**lithostatic** or **load pressure**), it is easy to deduce that xenoliths containing plagioclase come from depths ranging from a few metres down to 35 km, those with spinel come from 35–70 km, and those with garnet and diamond from even greater depths. More accurate information on the source of xenoliths can be obtained by detailed chemical and structural studies of individual minerals; but this is a specialists' subject well beyond the scope of this book.

The numerous studies carried out on xenoliths from various volcanoes around the world have shown that they mostly come from relatively shallow depths, with a maximum of 100–200 km. Xenoliths are, therefore, messengers that bring information about the outermost shells of the Earth.

Further evidence on the composition and structure of the Earth's interior is provided by **meteorites**. These are extra-terrestrial bodies that range from metallic (Fe–Ni) to silicate or mixed (metals plus silicates) in character. Some meteorites represent fragments of planets similar to the Earth that disintegrated due to collisions during the early life stages of the solar system. Their study tells us much about the internal structure and compositions of these ancient bodies and, thereby, also of our planet. Evidence from meteorites is somewhat elusive, yet it is critical for placing constraints on Earth composition, as it will be discussed soon.

The third important source of knowledge comes from geophysics, particularly from the study of seismic waves that cross the Earth during earthquakes. Seismicity will be reviewed in Chap. 5. For the scope of the present discussion, it is enough to know that earthquakes occur when rocks undergo a sudden fracturing under the effect of stress. Like any rigid body that breaks, rocks emit vibrations or **seismic waves**, which originate at the focal point or **hypocentre**, propagate quickly in all directions across the body of the Earth, and return to the surface, where they can be registered by seismographs (Fig. 1.2).

There are two types of seismic waves originating from the hypocentres: **P-waves** (*Primae*) and **S-waves** (*Secundae*), characterised by different energy, speed, and propagation mechanisms. P-waves can pass through solid and

---

[3] The pascal (Pa) is the unit of measurement of pressure, recommended by the *Système International* (SI). One gigapascal (GPa) is one billion Pa. Pressure values are also given in atmospheres (1 atmosphere = 101,325 Pa) or in bars (1 bar = 100,000 Pa).

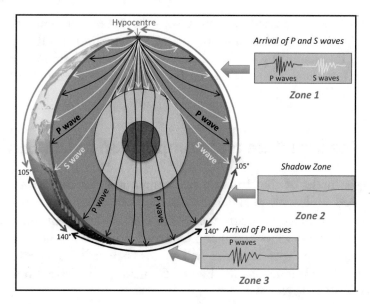

Hypocentre

Arrival of P and S waves

P waves   S waves

Zone 1

P wave
P wave
S wave
S wave
P wave
P wave

105°
105°
140°
140°

Shadow Zone

Zone 2

140° Arrival of P waves

P waves

Zone 3

Fig. 1.2 An earthquake generates energy that spreads across the body of the Earth as vibrations or seismic waves. There are two types of seismic waves originating at the hypocentres, the P-waves and S-waves. They follow curvilinear paths and can be recorded at remote monitoring stations by seismographs. Seismic stations situated between 0° and 105° from the hypocentre (Zone 1) record both the P-waves and S-waves. In the range between 105° and 140° (Zone 2 or Shadow Zone), neither P-waves nor S-waves are recorded because the former are refracted, and the latter are unable to pass through the Earth's liquid outer core. Stations located more than 140° from the hypocentre (Zone 3) only record P-waves, but not S-waves that are shielded by the outer liquid core

liquid materials, are fast and reach the recording stations first. S-waves are slower and, notably, cannot propagate through liquids.

The speed of P- and S-waves across the Earth (V_P and V_S) is variable, depending on the physical–chemical characteristics of the material through which they pass. Their values across minerals and rocks are well known, owing to the many experiments carried out in the laboratory at high pressures and temperatures. Comparison of experimental values with seismic wave velocities measured worldwide during earthquakes[4] allows us to determine what kind of material is present in the Earth's interior, between the hypocentre and the recording stations.

The paths of the seismic waves across the Earth are schematically outlined in Fig. 1.2. Stations located near the hypocentre only record waves that cross (and then can give information on) the shallower levels of the planet. In

---

[4] Values of V_P and V_S at a given locality are calculated by dividing the distance from the hypocentre by travel time, i.e. the time for seismic waves to reach the recording place.

contrast, more distant seismographs record the waves that penetrate deeper into the Earth. The composition of the materials resting at different depths is inferred by monitoring, at the seismic stations scattered around the world, the medium to large earthquakes that frequently occur at various sites of the planet.

1.3   The Structure and Composition of the Earth

Creating a model of the structure and composition of the Earth's interior has engaged philosophers and naturalists for centuries. The idea that the planet has a stratified structure was anticipated in the seventeenth century by René Descartes (1596–1650) and Gottfried Wilhelm von Leibniz (1646–1716), but a couple of centuries had to pass before being translated into a scientifically grounded model by the German geophysicist Emil Johann Wiechert (1861–1928).

A robust initial surge towards a scientifically supported model was made when Isaac Newton calculated that our planet has a mean density of about 5.51 g/cm^3, much higher than any rock occurring at the surface that rarely exceeds 2.7 g/cm^3. It was then obvious that the internal body of the Earth should consist of much heavier stuff. However, the very nature of these materials remained a matter of speculation.

Peridotite xenoliths and metal meteorites were the most plausible candidates because of their high density of about 3.3–3.5 g/cm^3 and 7–8 g/cm^3, respectively. Based on this evidence, geologists of the 19th and early twentieth century hypothesised that there were layers of peridotite and metals, under an external shell of relatively light rocks. The Austrian geologist Eduard Suess (1831–1914) gave the name of **Sial** to the outer rocky shell of the Earth, being composed of aluminium silicates; **Sima**, the intermediate layer, made up of more dense magnesium silicates; and **Nife** the nickel–iron core. A combination of silicates and metals was a brilliant solution to the density conundrum of the bulk Earth. However, the thickness, the state of aggregation (solid or liquid), and the exact composition of the individual layers remained obscure. Such problems had to be solved later in the twentieth century.

Integrated petrological[5] and geophysical investigations were decisive for working out a robust model of the interior of the Earth.[6] Seismic study

---

[5] Petrology (from the Greek words: πέτρος λόγος, pétros lógos = rock study) is the branch of geology that studies the texture, composition, physical characteristics and origin of rocks.
[6] Birch [1], Ringwood [2].

during earthquakes furnished values of V_P and V_S for materials residing at different depths; laboratory experiments provided critical data on the physical properties of minerals and rocks. A comparison of geophysical and petrological data obtained in the laboratories and in nature placed substantial constraints on the kind of materials within the Earth's interior and established the depth at which the transition from one type of material to the other occurred.

One of the most striking findings was that V_P and V_S changed with the depth, supporting the hypothesis of variable compositions for materials inside the Earth. Moreover, it was also discovered that there are some abrupt changes of seismic wave velocities at some particular depths (Fig. 1.3). These "jumps" are referred to as **seismic discontinuities** and were correctly interpreted as marking the boundary between layers with different mineralogical and/or chemical compositions, and/or distinct state of aggregation (solid–liquid).

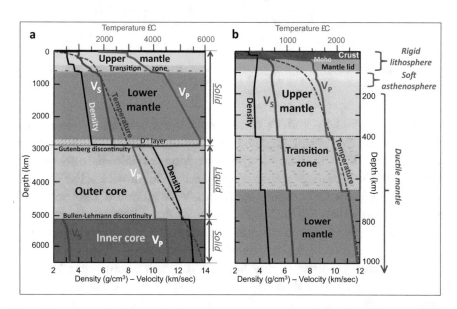

Fig. 1.3 a Simplified model of the internal structure of the Earth[7]; **b** Model of the crust-mantle system to 1000 km depth. Lines indicate the variation of seismic wave velocities, density, and temperature with depth. Density stratification is the consequence of the tendency of heavy material to sink below and the lighter one to float. Temperature variation with depth—called the **geothermal gradient**—is much higher in the crust (on average 30 °C/km) than in the mantle and core, reducing to about 1 °C/km in the centre of the Earth

[7] Dziewonski and Anderson [3].

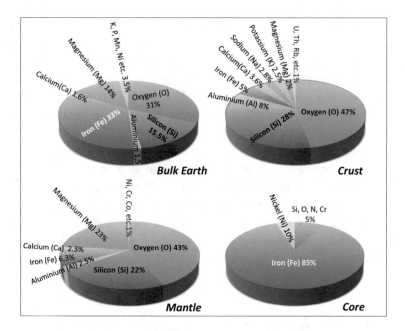

Fig. 1.4 Chemical composition of the bulk Earth, crust, mantle, and core. Element concentrations are expressed as per cent (%) by weight

The first jump of V_P and V_S occurs about 8 km beneath the ocean floor and at very variable depth (about 10 to 70 km; average 35 km) beneath the continents. This boundary is called the **Mohorovičić discontinuity** or **Moho**, after the Croatian geophysicist Andrija Mohorovičić (1857–1936), who discovered it in 1909. Above the Moho, V_P is around 6.5 km/sec, but it increases to more than 8.0 km/sec below the Moho. These values are consistent with a transition from aluminium silicate rocks in the upper domain to magnesium silicate rocks similar to the peridotite xenoliths below the Moho.

A much stronger discontinuity lies at a depth of about 2900 km. It is known as the **Gutenberg discontinuity**, named after the German-American geophysicist Beno Gutenberg (1889–1960), who discovered it in 1913. Below this discontinuity, a strong decrease of V_P and a loss of S-waves ($V_S = 0$) highlights the most dramatic change in the structure and composition of the Earth's interior, i.e. the transition from a solid rocky upper layer to a liquid metallic mass through which the P-wave velocities are strongly retarded, and S-waves are not allowed to pass (Fig. 1.2).

A third major discontinuity is present at about 5150 km, where V_P increases from about 10 to 11 km/sec. This is the **Bullen-Lehmann discontinuity**, named after the Danish seismologist Inge Lehmann (1888–1993), who hypothesised its existence in 1936, and New Zealander geophysicist Keith Edward Bullen (1906–1976), a prominent scholar of seismology.

The studies summarised above resulted in the recognition of various compositionally distinct concentric spherical layers or shells inside the Earth, as Suess had suggested in the late nineteenth century. However, the situation is somewhat more complex. In essence, three layers—crust, mantle and core—can be distinguished from a chemical point of view; by contrast, when mechanical characteristics are considered at least four different shells are recognised—the inner core (solid), the outer core (liquid), the convective mantle (solid ductile) and the lithosphere (solid rigid). Mechanical and chemical layers do not match and both types are to be considered to explain many first-order geological processes such as geomagnetism, seismicity, ocean basin formation, and mountain building.

1.3.1 The Compositional Layering: Core, Mantle, and Crust

The **core** extends from the centre of the Earth—situated at a depth ranging from 6357 km at the poles and 6378 km at the equator—to the Gutenberg discontinuity. It consists of an inner sphere of solid metal enveloped by a thick liquid layer, separated by the Bullen-Lehmann discontinuity. Core composition is dominated by iron, with lesser amounts of nickel (the so-called *Nife* of Suess) and small contents of light elements such as silicon, nitrogen, sulphur and water (Fig. 1.4).

The **mantle** is placed between the Gutenberg discontinuity and the Moho. It is subdivided into three distinct concentric layers by second-order seismic discontinuities: the **upper mantle** (extending down to 400 km), a **transition zone** (400–650 km), and the **lower mantle** (650–2900 km). A distinct irregular domain, a few hundred kilometres thick named the **D" layer** (D-double-prime layer), is located at the core-mantle boundary. The mantle is believed to have a relatively homogeneous chemical composition dominated by oxygen, silicon, and magnesium that resembles the *Sima* of Suess. However, mineralogical composition changes vertically, as a response to the progressive increase in pressure and temperature with depth. Seismology and experimental petrology suggest that the upper mantle is formed of peridotite. By contrast, the transition zone and the lower mantle are made up of some exotic high-density minerals that are stable at very high pressures,

Table 1.1 Mass, volume, and density of the Earth and its main structural units

	Mass (Tons)	Volume (km^3)	Percent of Earth mass	Percent of Earth volume	Density (g/cm^3)
Bulk Solid Earth	5.97 × 10^{21}	1.13 × 10^{12}	100	100	5.5
Atmosphere*	5.10 × 10^{15}	2.6 × 10^{10}	0.00008	2.3	0–1.2 × 10^{-3}
Hydrosphere	1.40 × 10^{18}	1.40 × 10^9	0.023	0.12	1.03
Bulk Crust	2.59 × 10^{19}	9.25 × 10^9	0.43	0.09	2.8
Oceanic Crust	5.91 × 10^{18}	2.0 × 10^9	0.1	0.04	2.9
Continental Crust	20 × 10^{18}	7.25 × 10^9	0.33	0.05	2.7
Mantle	4.06 × 10^{21}	9.5 × 10^{11}	68.05	84.15	3.3–6.0
Bulk Core	1.89 × 10^{21}	1.78 × 10^{11}	31.5	15.76	11
Outer Core	1.79 × 10^{21}	1.7 × 10^{11}	30	15	10
Inner Core	9.7 × 10^{19}	7.6 × 10^9	1.5	0.7	13

*Troposphere + stratosphere

such as wadsleyite, ringwoodite, Fe-periclase, perovskite, Mg-Si-perovskite, and Mg-wüstite.[8]

The **crust** is the external rocky shell extending from the surface of the solid Earth down to the Moho. Its thickness is variable, ranging from an average of about 8 km under the ocean basins to a maximum of about 70 km under some mountain chains. The composition and density of the crust change significantly from oceans to continents (Table 1.1). The **oceanic crust** is made up of a dense layer of igneous rocks (basalt and gabbro; Box 1.1) covered by a veneer of sediments; this rock suite extends rather uniformly over vast areas, flooring the Earth's oceans. Its chemical composition consists predominantly of oxygen, silicon, aluminium, and calcium with considerable iron, hosted in a small number of minerals, namely plagioclase feldspar, pyroxene, olivine and Fe-oxides.

---

[8] Wadsleyite and ringwoodite make up the transition zone; they have the same composition as olivine $(Mg,Fe)_2SiO_4$ but atoms are more closely packed, resulting in a higher density; phases of the lower mantle are Fe-periclase $(Mg,Fe)O$, Mg-Si-perovskite $(Mg,Fe)Al_2SiO_6$, perovskite and post-perovskite $CaSiO_3$, and Mg-wüstite $(Mg,Fe)O$.

The **continental crust** is highly heterogeneous, being composed of a large number of igneous, sedimentary, and metamorphic rocks (see Box 6.2). Its chemical composition is also enriched in oxygen, silicon and aluminium, but calcium and iron are lower, and silicon is higher than in the oceanic crust. Main constituent minerals include silicates and carbonates such as quartz, feldspars, and calcite. These minerals have a much lower density than those making up the oceanic crust, which makes the continental crust much lighter than under the oceans.

The structure described above is not an intrinsic characteristic of the planet but has instead been acquired slowly over geological time, mainly as a consequence of magmatic processes, which have been operating continuously within the Earth since shortly after its formation 4.56 billion years ago and are still going on at present.

As will be discussed in the following chapters, the individual layers are not static but exchange energy and matter, making the Earth a continuously evolving, restless planet. The continental crust has gradually increased in volume from the small protocontinents of the Archaean (4.0 to 2.5 billion years ago) to the great continental masses of the Phanerozoic (~540 million years ago to present). The mantle has cooled down slowly over time (about 50 °C every billion years) and has gradually lost some of its original chemical components, such as alkali metals, phosphorous, uranium, thorium, rare earth elements (REE), nitrogen and water, which have been progressively concentrated in the crust and the fluid Earth by magmatic processes (Chap. 3). The core was formed early in Earth's history, but began to solidify around 1.5 billion years ago. Its solid mass has grown through time at the expense of the liquid outer core. Crystallisation has released heat (**latent heat**) and chemical impurities that have migrated to the liquid outer core, sustaining thermal and compositional convection.

1.3.2 The Mechanical Layering: Inner Core, Outer Core, Convective Mantle, and Lithosphere

The mechanical layering of the Earth does not match with chemical layering. Going into some detail, the core has a rather uniform chemical composition dominated by Fe–Ni metal. However, the inner core is a stiff solid sphere of crystallised metal, whereas the outer core is liquid and is subjected to turbulence. Movements of the liquid metal within the outer core cause the Earth's magnetic field, a feature that is indispensable to life (Chap. 4). Simply put, the mechanical layering of an otherwise compositionally homogeneous core is the prime reason for geomagnetism.

The largest portion of the mantle—the lower mantle, the transition zone, and a large part of the upper mantle—consists of ductile rocks that easily deform when subjected to oriented stresses, in the same way slightly heated wax deforms under pressure from a finger. Because of these properties and the internal temperature gradients (Fig. 1.3), the ductile mantle undergoes slow continuous convection movement at a rate of a few cm/year; therefore, it is also termed the **convective mantle**. Mantle convection provides a very efficacious mechanism of heat transfer toward the planet's external layers.

The uppermost ductile mantle grades upward into a shell of very pliant peridotite, characterised by particularly "soft" mechanical properties and diminished seismic wave velocities. This is called **asthenosphere** (from the Greek word ἀσθενής, asthenés = weak) or **low-velocity channel** and separates the convective mantle from the overlying rigid lithosphere. The soft mechanical characteristics of the asthenosphere result from the presence of small amounts of magmatic liquid dispersed in the interstices of peridotite-forming minerals.

Finally, the **lithosphere**[9] is the outermost mechanical layer of the solid Earth. It comprises a package made of the mantle lid, plus a superimposed oceanic or continental crust. The **lid** is made of peridotite rocks similar to the rest of the upper mantle; however, the temperature is lower, which makes the lid's mechanical properties rigid. The mantle lid is tightly attached to the overlying crust and together form a discrete and mechanically coherent package that is relatively independent from the convective mantle.

The mantle lid is a common substratum of the lithosphere. However, whereas a thin and dense basaltic oceanic crust overlies the mantle lid in the oceanic lithosphere, the continental lithosphere consists of the lid plus a thick but relatively low-density continental crust (Fig. 1.5). As a result, the oceanic lithosphere is thinner but denser than the continental counterpart. The variable thickness of the lithosphere in the oceanic and continental environments have strong effects on the mantle melting and the origin of magmas inside the Earth.[10]

The thickness of the oceanic lithosphere is fairly constant, in the order of 100 km. Because of its high density, the oceanic lithosphere uniformly lays down over the underlying asthenosphere, forming depressed basins covered by oceans. The thickness of the continental lithosphere is much higher,

---

[9] The term lithosphere is sometimes mistakenly used as a synonym for crust. However, the crust is only one layer of the lithosphere. The lithosphere (mantle lid + crust) is defined by its rigid mechanical properties, whereas the crust is defined on the basis of its chemical and mineralogical composition. In short, the lithosphere consists of two compositionally distinct layers, but behaves as a single mechanically coherent unit.

[10] Niu [4].

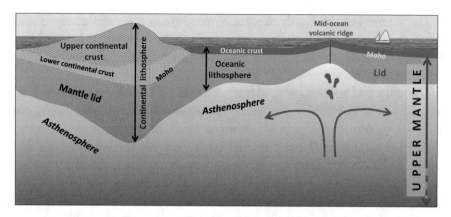

Fig. 1.5 Idealised section of the Earth's lithosphere and upper mantle. The oceanic lithosphere consists of the mantle lid and a thin but dense basaltic crust. The continental lithosphere contains a thick but low-density continental crust. The continental lithosphere rises by floating (isostasy) on the underlying asthenosphere to form the landmasses. The convective mantle is active beneath the lithosphere (red arrows) and breaks it apart along continental rift valleys and submarine fracture zones marked by long volcanic chains known as mid-ocean ridges

reaching about 250–300 km under some mountain chains and in geologically old and stable regions (**cratons**). The overall density of the continental lithosphere is low and, for this reason, sinks moderately into the underlying asthenosphere, like the hull of a ship in the water, receiving an upward thrust by buoyancy (**isostasy**). Isostatic floating makes the continents rise above sea level to create the emergent land upon which all of us live.

Because of its stiff mechanical nature, the lithosphere breaks up into a large number of slabs called **lithospheric plates**. Plates are mobile and interact across the Earth's surface, floating like rafts on the underlying asthenosphere. The mobility of the lithospheric plates is the cause of seismicity, volcanism, the building of mountain belts, the formation of oceans, and numerous other geological phenomena (Chap. 6). However, plate mobility is made possible by the asthenosphere that forms a kind of 'lubricated' shell that separates the convective mantle from the lithospheric plates.

1.4 Summary

The structure and composition of the Earth's interior is revealed by studies of rock fragments brought to the surface by volcanic eruptions (xenoliths), meteorites, and the behaviour of seismic waves that cross the planet during earthquakes.

The internal of the Earth consists of three main compositionally distinct layers: the crust, mantle, and core, also indicated by geologists historically as Sial (i.e. made of aluminium silicates), Sima (magnesium silicates) and Nife (nickel and iron), respectively.

The crust has an average thickness of about 35 km of the continents and about 8 km of the oceans. It consists essentially of silicate rocks, containing feldspars, quartz, and Fe–Mg-Ca minerals. Si-Al minerals, such as quartz and felspars, dominate in the continental crust, whereas the Fe–Mg-Ca silicate minerals olivine, pyroxene and plagioclase make up the rocks of the oceanic crust. The contrasting mineralogical composition makes the oceanic crust density much greater than that of the continental crust.

The mantle extends from the base of the crust to a depth of 2900 km and consists of peridotite (olivine and pyroxenes) up to about 400 km and, at deeper levels, by high-density magnesium, calcium silicates, and iron oxides.

The core consists of iron, with some nickel and minor amounts of light elements such as oxygen, nitrogen, and/or silicon.

The boundary separating the crust and the mantle is known as the Mohorovičić discontinuity or Moho. The transition between the mantle and the outer core is called Gutenberg discontinuity and separates the silicate from the metallic Earth. The Bullen-Lehmann's discontinuity divides the outer liquid core from the inner solid core.

The uppermost layer of the mantle, called the mantle lid, is tightly joined with the overlying crust. Together, these form a single rigid layer called the lithosphere. The lithosphere is divided into several large slabs or plates that move horizontally on the Earth's surface to produce geologically important processes, such as the formation of mountain chains, the opening of oceans, seismicity, and volcanism. The separating level between the lithospheric plates and the underlying mantle is represented by the asthenosphere—a layer of peridotite showing very soft mechanical characteristics due to the presence of tiny amounts of magmatic liquid.

The crust, mantle, and core have evolved over geological time, and their physical and chemical characteristics have been continuously changing. The solid internal core is crystallising at a rate of about 1 mm/year, delivering chemical impurities and latent heat of crystallisation to the outer liquid core, whose volume has slowly reduced with time. The outer liquid core is subject to rapid movements, which are responsible for the formation of the Earth's magnetic field. The mantle is presently much cooler than it was in the early stages of Earth history and, over time, has become deficient in particular chemical elements (e.g. alkali metals, U, Th, rare earths, and water) that have been extracted by magmas and transferred upwards to make up the crust and the external fluid spheres of the Earth.

1.5 Box 1.1—Minerals and Rocks

The Earth, the other inner planets of the solar system, and the asteroids are made up of rocks, an aggregation of one or more minerals.

Minerals are natural solid materials with particular physical and chemical properties. Their structure is characterised by an ordered arrangement of the constituent atoms, which replicates indefinitely in space (**crystal lattice**). The smallest part of any mineral or other crystalline substances is the **unit cell** (Fig. 1.6a), which represents the composition and structure of the entire crystal.

Minerals are sometimes found as crystals with well-developed geometrical forms and remarkable sizes (Fig. 1.6b), sometimes reaching several meters; in rocks, they occur as tiny grains that are best observed with a magnifying lens or under a microscope (Fig. 1.6c).

There are several classes of minerals, such as the native elements, silicates, carbonates, sulphates, sulphides, oxides, hydroxides, and phosphates.

Native element minerals occurring in nature are rare but include many interesting species, in particular graphite and diamond—two different forms, or polymorphs, of carbon stable at different pressures—copper, gold, and many others.

Silicates make up most of the crust and upper mantle, Common silicate minerals include plagioclase feldspar (a mixture of albite [$NaAlSi_3O_8$] and anorthite [$CaAl_2Si_2O_8$]), quartz [SiO_2], alkali-feldspars [$(Na,K)AlSi_3O_8$],

Fig. 1.6 a Unit cell of quartz (SiO_2); volume is around 0.2–0.3 cubic nanometres; **b** Quartz crystals; **c** Granules of quartz and feldspars in a sedimentary rock observed under the microscope

pyroxenes [$(Ca,Fe,Mg)_2Si_2O_6$], amphiboles (complex string silicates consisting of Fe, Mg, Ca, Al, and Ti), olivine [$(Mg,Fe)_2SiO_4$], mica (sheet silicates with complex formulae containing K, Na, Li, Al, Fe, Mg, OH, and F), clay minerals (hydrated silicates of variable composition) and some species such as zircon [$ZrSiO_4$], occurring in small amounts as an **accessory mineral** in many rocks.

Carbonates are abundant within the crust. The most abundant are calcite, aragonite [$CaCO_3$], and dolomite [$CaMg(CO_3)_2$]; calcite and aragonite have the same chemical formula, but have a different crystal structure and physical characteristics.

Sulphate minerals include gypsum [$CaSO_4 \cdot 2H_2O$], anhydrite [$CaSO_4$], and barite ($BaSO_4$] and many others.

Among sulphides, pyrite [FeS_2] is common; oxides and hydroxides, such as magnetite [Fe_3O_4], hematite [Fe_2O_3], goethite [$FeO(OH)$], limonite, pyrolusite [MnO_2], are widespread.

Apatite [$Ca_5(PO_4)_3(OH,F,Cl)$] and zircon [$ZrSiO_4$] are common accessory phosphate and silicate minerals. Monazite [$(Ce,La)PO_4$] is much rarer.

Rocks are natural solid objects consisting of an aggregate of one or more mineral granules. **Silicate rocks** are made up prevailingly of silicate minerals. **Carbonate rocks** consist primarily of calcite or dolomite. Rocks are divided into three large families: magmatic, sedimentary, and metamorphic. They are genetically separated but can convert one into the other during the **Rock Cycle** (Box 7.1).

Magmatic or **igneous rocks** form by the solidification of magmas either inside the crust (**intrusive rocks**), or on Earth's surface after eruption (**volcanic** or **effusive rocks**). Intrusive and effusive rocks that originate from the same magma have equal chemical composition but exhibit different texture, i.e. the geometric features of the component minerals. Intrusive rocks are made of centimetre sized mineral granules, whereas effusive rocks consist of fine-grained minerals, sometimes surrounding larger crystals referred to as phenocrysts (**porphyritic texture**). Some volcanic rocks consist largely or entirely of glass (e.g. obsidian, pumice). Igneous rocks have a silicate composition, except for very rare and exotic types composed of carbonate (**carbonatites**).

Sedimentary rocks are formed by the accumulation and lithification of fragments of pre-existing rocks (**clastic rock**), remains of living organisms (**biogenic rocks**), or minerals precipitated from oversaturated aqueous solutions (**chemical rocks**). Sedimentary rocks are generally stratified and cover the largest area of the Earth's surface (70–80%), although their mass represents a mere 7–8% of the crust. Constituent minerals are mainly silicates (e.g.

in sandstones and clays), carbonates (limestone, dolostone) and, to a much lesser extent, sulphates, oxide-hydroxide etc.

Metamorphic rocks are the products of metamorphism, a process of mineralogical and structural modification that rocks of any kind undergo when subjected to high temperatures and pressures inside the Earth. Under such conditions, the original rocks become unstable and change their mineralogical composition and structure (**metamorphic recrystallisation**). However, the original chemical composition can be preserved during metamorphism, although, in some cases, it is modified by interaction with fluids (**metasomatism**). Metamorphic recrystallisation takes place at temperatures typically in the range 150–650 °C. At higher temperatures, the rocks undergo partial melting (**anatexis**), generating magmas. Metamorphic rocks are entirely made up of crystals and, therefore, are also referred to as **crystalline rocks**.

The names and compositions of the main rock types are as follows:

Igneous rocks

Andesite: Volcanic rock composed of plagioclase feldspar, pyroxene, and hornblende.

Basalt: Effusive rock formed of pyroxene, plagioclase feldspar, and olivine.

Diorite: Intrusive rock with the same composition as andesite.

Gabbro : Intrusive rock equivalent to basalt.

Granite: Intrusive rock composed of quartz, feldspars, and biotite. It makes up large sectors of the continental crust.

Rhyolite: Effusive equivalent of granite.

Sedimentary rocks

Chalk: Soft, friable, white variety of carbonate rock formed by the accumulation of microscopic plankton shells.

Chert: Chemical or biogenic sedimentary rock consisting of silica (SiO_2) such as quartz and opal.

Claystone: Very fine-grained clastic or chemical sedimentary rock formed of clay minerals.

Dolostone: Carbonate rock mainly made up of dolomite.

Limestone: Chemical, organogenic, or clastic sedimentary rock formed predominantly of calcite.

Marl or *marlstone*: fine-grained carbonate rock containing variable amounts of clays and silt (35-65%).

Sandstone: Clastic rock generated by the lithification of sandy debris derived from the breakdown of other rocks. Sandstones typically consist of quartz and feldspar grains held together by an intergranular clay or carbonate cement.

Travertine: Chemical sedimentary rock formed by the precipitation of calcite from thermal waters.

Metamorphic rocks

Eclogite: High-pressure, high-density rock made up of green pyroxene (omphacite) and garnet.

Gneiss: A medium- to coarse-grained banded metamorphic rock that is rich in feldspars and quartz, but also contains mica minerals and aluminous or Ca-Fe-Mg silicates (e.g. garnets).

Granulite: High-temperature metamorphic rock consisting of feldspars, quartz, pyroxenes, and garnet. It occurs at deep levels in the continental crust.

Marble: Carbonate metamorphic rock with a granular texture, derived from limestone or dolostone.

Peridotite: Metamorphic rock composed of granules of olivine and pyroxenes. It can also be generated by magmatic processes.

Phyllite: Fine-grained rock consisting of quartz, mica, and feldspars having an oriented texture (cleavage).

Schist: Medium-grained rock consisting of quartz, mica, and feldspars with a well-developed oriented texture (schistosity), resulting from the parallel arrangement of mica flakes.

Shale: Very fine-grained metamorphic rock formed of mica, quartz, and feldspar having an oriented texture.

References

1. Birch F (1952) Elasticity and constitution of the Earth's interior. J Geophys Res 57:227–286
2. Ringwood AE (1975) Composition and Petrology of the Earth's Mantle. McGraw-Hill, 618 p
3. Dziewonski AM., Anderson DL (1981) Preliminary reference Earth model. Phys Earth Planet Inter 25:297–356
4. Niu Y (2021) Lithosphere thickness controls the extent of mantle melting, depth of melt extraction and basalt compositions in all tectonic settings on Earth - A review and new perspectives. Earth-Sci Rev 217:1–25

2

Air, Water, Earth—The Exogenic Geological Processes

*Nel cuore delle pietre è il seme, il polline, il fiore la foglia disseccata, il ramo secco,
ossificate radici; l'ape, il grillo, la fragile farfalla..... Nel cuore delle pietre giace la
goccia disseccata, laghi prosciugati e paludi, il flusso riarso delle sorgenti, l'afona voce
degli oceani, i pesci e le meduse, alghe azzurro-verdi ... (In the heart of stones is the
seed, the pollen, the flower, the dried leaf, the dead tree branch, ossified roots; the bee,
the cricket, the graceful butterfly . . . In the heart of the stones lie the vanished drop,
dried-up lakes and marshes, the parched flow of springs, the hoarse voice of the oceans,
fish and jellyfish, blue-green algae . . Marcello Feola, The Hearth of Stones.)*
Marcello Feola, contemporary Italian poet - Il cuore delle pietre

2.1   Introduction

The best-known geological events occur on the Earth's surface. These are
referred to as **exogenic** or **supergene** and involve the interaction of atmosphere, water and rocks, the spheres that ancient scholars identified as *Air,
Water*, and *Earth*. The supergene geological processes can be framed into the
broader perspective of the so-called **sedimentary cycle**, a set of phenomena
that begin with the weathering of rocks and conclude with deposition and
compaction of sediments and, ultimately, the formation of new rocks.

© The Author(s), under exclusive license to Springer Nature
Switzerland AG 2021, corrected publication 2022
A. Peccerillo, *Air, Water, Earth, Fire*,
https://doi.org/10.1007/978-3-030-78013-5_2

The moulding of landforms is the most visible manifestation of the sedimentary cycle. But there are other important effects, such as the formation of soils, the modification of the composition of the atmosphere and the hydrosphere, and the accumulation of fossil fuels. These aspects are less known but have a critical impact on the natural environment.

Since the sedimentary cycle occurs in the outermost spheres of the Earth, it is necessary to have a quick look at the physicochemical characteristics of soils, rocks, hydrosphere, and the atmosphere—the main systems through which exogenic processes progress.

2.2   Rocks and Soils

Most of the Earth's crust is made up of metamorphic and igneous rocks. Sediments are a small fraction of the total rock mass, but they cover some 70–80% of the land surface and almost all of the seafloor. Metamorphic and igneous rocks are largely made of silicate minerals, whereas sedimentary rocks typically consist of silicates, carbonates, especially calcite and dolomite (Box 1.1), and mixtures of the two.

Most rock-forming minerals are unstable under the chemical and physical conditions of the Earth surface. Therefore, they tend to react with water and atmospheric gases to release elements into waters and form new minerals that are stable in the supergene environment (Box 2.1). **Soil**, or the **pedosphere,** is the product of such chemical modification, although living organisms significantly assist in its formation and subsequent development.

Soils are a mixture of rock fragments, mineral grains and their alteration products, organic matter, gases, and water. They are commonly present as an unconsolidated layer that extends over large areas of the land surface. Soils sustain plant growth and are the habitat for many living organisms. They also host significant amounts of organic and inorganic carbon and are one of the largest exogenic reservoirs of this element on Earth (Chap. 7). The frozen soil layers present at high latitudes, known as **permafrost**, are estimated to contain billions of tons of carbon, which will likely be released into the atmosphere by global warming.

Soils have existed since at least the Archaean aeon (4.0–2.5 billion years ago), although their formation mechanisms and composition have changed over time. The remains of ancient soils (**palaeosols**) are commonly found within continental geological sequences and are invaluable sources of information about the environmental conditions in the past.

Rocks and the pedosphere dramatically affect the composition of the atmosphere and hydrosphere. They also provide nutrients to living organisms and are crucial for the origin and growth of the **biosphere**, both on the continents and in the oceans. Some elements essential to life, such as potassium, phosphorus and iron, derive primarily from rocks and soils. Without these precious resources, there would be no flow of nutrient elements to the sea, which would only be an enormous infertile water mass.

Therefore, rocks are not the arid objects they might appear to be at first sight, but instead are primary suppliers of nourishment for living organisms. From the microscopic marine plankton to the tallest trees, the entire Earth's biosphere derives its sustenance from rocks: a truism well known not only by geologists and pedologists, but also to poets, as demonstrated by verses quoted at the beginning of this chapter.

2.3   The Atmosphere

The atmosphere has a stratified structure, much like the rest of the planet. The troposphere is the lowermost layer extending from the surface to an altitude of about 12 km. Above this, are the stratosphere (up to about 50 km), the mesosphere (50–85 km), the thermosphere (85–690 km), and then the outermost region of the planet's atmosphere known as the exosphere. About 80% of the gaseous mass and almost all the water vapour are contained in the troposphere; the remaining 20% is essentially hosted in the stratosphere.

The atmosphere consists primarily of molecular nitrogen and oxygen (N_2=78% and O_2=21% by volume), plus minor to trace amounts of other gaseous species such as the noble gases (especially argon, Ar=0.93%), carbon dioxide (CO_2=0.04%), and traces of carbon monoxide (CO), methane (CH_4), ozone (O_3), hydrogen (H_2), and nitrogen oxides (NO_x). The content of water vapour in the atmosphere is variable, ranging from near zero to 4–5% by volume.

There has always been an atmosphere around the Earth. Most of its mass derives from the early separation of gaseous substances from the crust and mantle during their solidification about 4.5 billion years ago (Chap. 3). Subsequent continuous release through volcanism and diffuse degassing along deep crustal fractures or faults gave substantial contributions.

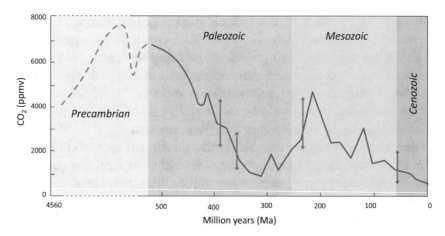

Fig. 2.1 Simplified variation of CO_2 in the atmosphere from Precambrian to present. The concentration of CO_2 is expressed in parts per million by volume (ppmv), which is the volume of carbon dioxide contained in one million parts of atmospheric gases. The curve is the average of data furnished by proxies, i.e. the geological materials whose properties can be measured and correlated with atmospheric compositions and climatic parameters (see Chap. 8). The vertical bars are uncertainties in the estimates[1]

The Earth's primordial atmosphere had an extremely different composition than at present. Nitrogen was the main component, but hydrogen and various reduced species, such as methane (CH_4), ammonia (NH_3) and hydrogen sulphide (H_2S), were in high abundance. Carbon dioxide (CO_2) was initially present at moderate concentration, but afterwards expanded to some 1–2% or higher, and then progressively declined over time to its present concentration of a few hundred parts per million (Fig. 2.1). Oxygen was virtually absent in the Earth's early atmosphere because of reaction with reduced compounds and atoms that quickly consumed the modest quantities formed by dissociation of water molecules under the effect of ultraviolet radiation (**photolysis**).[2] An increase in the volume of atmospheric oxygen to 2–3% occurred between 2.4 and 2.0 billion years ago, during the so-called **Great Oxidation Event** (Chap. 9). Such a build-up originated from the appearance and expansion of single-cell organisms capable of photosynthesis (**cyanobacteria**) but with substantial help from geological events, in particular changes in the composition of volcanic gases and magmas. Modification in volcanism resulted from a revolution in the large-scale movements

---

[1] Royer et al. [1].

[2] Photolysis is the chemical process by which molecules of water, or other compounds, are broken down into their component atoms through the absorption of light.

Table 2.1 Volumes of the hydrosphere reservoirs (in millions km³)

Reservoir	Volume $\times 10^6$ km³	Per cent of total hydrosphere
Oceans	1370	97.3
Ice caps and glaciers	29	2.03
Deep groundwater	5.3	0.37
Shallow groundwater	4.3	0.3
Lakes	0.125	0.01
Soil moisture	0.065	0.005
Atmosphere	0.013	0.001
Rivers	0.0017	0.0001
Biosphere	0.001	0.00007
Total	*1409*	*100*

of the Earth's crust and lithosphere (global tectonics) that occurred during the Mesoarchean, about 3.2 billion years ago.

In addition to biological activity and volcanism, the composition of the atmosphere has been profoundly modified by chemical reactions with rock minerals, which consumed huge amounts of some components, especially CO_2 (Box 2.1).

2.4   The Hydrosphere

The Earth's water is contained in the oceans, seas, lakes, rivers, ice caps and glaciers, and as moisture in soils and the atmosphere; considerable amounts are stored in the pores and fractures of rocks, forming underground wet layers known as **aquifers**. A very tiny but vital amount of water is contained in the biosphere (Table 2.1).

Seawater comprises 97.3% of the total hydrosphere, the rest being types of freshwater. Its physical and chemical characteristics are extremely variable.[3] Rainwater contains significant amounts of dissolved atmospheric gases and impurities whose concentrations decrease inland from the coast and increase from rural to industrial areas (Table 2.2). River and groundwater compositions depend mainly on host rocks and soils, as well as the time spent in contact with them. An important and sometimes dramatic role in

---

[3] Krauskopf and Bird [2].

Table 2.2 Typical composition of natural waters expressed in parts per millions (ppm; milligrams of chemical components per litre of solution)[4]

Chemical component	River water mg/l (ppm)	Rain water Amazon (ppm)	Rain water Hawaii (ppm)	Sea water (ppm)
HCO_3^-	55.9	trace	trace	137
SO_4^{2-}	10.6	0.28	1.92	2717
NO_3^-	8.1	0.056	0.2	0.18
Cl^-	-	0.39	9.63	19,010
Ca^{2+}	3.9	0.008	0.47	413
Mg^{2+}	3.9	0.012	0.92	1296
Na^+	6.9	0.23	5.46	10,800
K^+	2.1	0.012	0.37	407
SiO_2	13.1	trace	trace	0.5–10
Total	*105*	*0.99*	*18.97*	*34,785*
pH	*5.3*	*5.3*	*4.8*	*8.1*

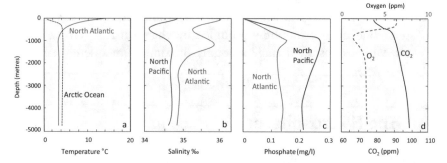

Fig. 2.2 Vertical variations of some physical and chemical parameters in the ocean water. The positive spike of salinity at about 1000 m depth in the North Atlantic is an effect of the inflow of salty water from the Mediterranean Sea through the Strait of Gibraltar

water chemistry occurs with the introduction of external material related to biological and human activities.

The physical and chemical characteristics of marine waters (ocean and salty seas) change strongly both laterally and vertically. Except in the polar regions, the **temperature** decreases sharply in the first hundred meters from the surface and then remains almost constant with the depth. At the bottom of all oceans, therefore, the temperature is relatively low—close to zero degrees Celsius—even at the equator (Fig. 2.2a).

---

[4] One part per million (ppm) denotes one part of a chemical component per 1 million parts of solution or rock, or else. One ppm is equivalent to 1 mg of solute per litre of water (mg/l) or 1 mg of element per kilogram of rock or soil (mg/kg).

The **density** of pure water is slightly less than 1 g/cm^3 at 20 °C, and decreases with temperature. Values in the oceans range from about 1.020 to 1.029 g/cm^3 the higher values being characteristic of cooler and more saline waters. These small density variations are crucial for oceanic circulation. Dense waters sink deep towards the ocean floor, spread out horizontally on the bottom, return to the surface and move horizontally as surface currents, to replace sunken heavy water and close the circulation loop. The sinking of dense water occurs at a few places around the globe, namely in the Weddell Sea and the Ross Sea in the Antarctic, and near southeast Greenland in the Arctic. Water density at polar latitudes is high because of both low tempera-tures and salinity. High salt contents are related to freezing, since ice is unable to incorporate salts, which concentrate in the liquid water.

The global ocean water flux is called **thermohaline circulation** because it is triggered by density contrasts driven by temperature and salinity. Thanks to deep circulation, there is a continuous supply of oxygen to the deeper ocean waters.

The **chemical composition** of seawater is comprised of ions in solu-tion, dissolved gases (oxygen, carbon dioxide, nitrogen, and others), and suspended inorganic and organic matter, including countless types of bacteria and viruses. The largest fraction of chemical components is transported from the mainland to the ocean by rivers or comes directly from the atmosphere; a minor amount is provided by submarine volcanic and hydrothermal activity, plus by seawater reacting chemically with rocks of the seafloor. The ocean water contains the same type of solutes as freshwater, but absolute concen-trations are much higher, and relative abundances of chemical species are different (Table 2.2). These differences are related to several factors, such as the selective concentration in the seawater of the most soluble ions (e.g. Na$^+$, Cl$^-$) and the biological activity of living organisms that consume higher amounts of some elements with respect to others (e.g. Ca *vs* Mg) during their life cycle.

Salinity is the total amount of salt dissolved in water. Ocean water has a mean value of about 35 parts per thousand by weight (grams per litre of water typically denoted as ‰), whereas freshwater generally has salinities lower than about 0.5‰. The highest salt concentrations are found in the closed seas such as the Mediterranean Sea (about 37‰) and the Red Sea (about 36–41‰), and in some closed lakes strongly affected by evaporation whose salinity may exceed 50‰ (**brines**). Low salinity is typical of basins where the inflow of freshwater largely prevails over evaporation, such as in the Black Sea (about 20‰) and the Baltic Sea (between about 2–10‰). In most cases, salinity is higher in surface waters because of evaporation (Fig. 2.2b).

The most abundant ions in the seawater are chlorine and sodium ($Cl^- = 1.9\%$; $Na^+ = 1.1\%$ by weight, on average), followed by sulphate (SO_4^{2-}), magnesium, calcium, potassium, and bicarbonate (HCO_3^-). Silicon, aluminium, phosphorous and iron are present in trace amounts due to their low solubility. Some of these elements are nutrients for phyto-plankton to synthesise molecules; therefore, their scarcity dramatically limits the development of marine life. Note, however, that the concentrations of nutrients, such as phosphorous, generally increase with depth due to the dissolution of the rests of organisms falling from above (Fig. 2.2c).

Gases dissolved in natural waters are the same as the atmosphere due to continuous exchanges between the two systems. The most abundant species, therefore, include nitrogen, oxygen, argon, and carbon dioxide. **Oxygen** occurs as dissolved diatomic molecules (O_2) in the range of a few parts per million (*ppm* = milligrams of oxygen per litre of water); its concen-tration in the ocean is higher at the surface and rapidly reduces with the depth, reaching a minimum at about a thousand meters, where oxygen is consumed by reaction with dead organic material falling from above. Succes-sively, oxygen increases slightly or remains constant down to the sea bottom (Fig. 2.2d). In small closed basins, where the vertical transfer of oxygen is modest or lacking (e.g. the Black Sea and many lakes), oxygen-poor anoxic subsurface conditions develop, making deep waters hostile to life.

Carbon dioxide (CO_2) is ubiquitous in natural waters. Its solubility decreases with temperature and increases with pressure. Therefore, its abun-dance is higher in the deep ocean than near the surface (Fig. 2.2d). At shallow-moderate depths, CO_2 can dissolve in a molecular form or react with water to produce hydrogen, bicarbonate and carbonate ions (**hydrolysis**). The greater the quantity of CO_2 and the higher the number of hydrogen ions, i.e. the stronger the acidity of the solution.[5] Temperature and pressure at the ocean bottoms cause CO_2 to turn into solid compounds, called **hydrates**, that become stable at depths of around 3600 m. Such a modification stabilises CO_2 in deep waters, making ocean bottom an appealing location for the possible storage of excess atmospheric CO_2 produced by fossil fuel burning[6] (see Chap. 8).

The **acidity** of natural water (or any other solution) is expressed by its *pH*, which is defined as the negative decimal logarithm of the dissolved hydrogen

---

[5] The reaction of CO_2 dissolution is: $CO_2 + H_2O \leftrightarrow H_2CO_3 \leftrightarrow H^+ + HCO_3^- \leftrightarrow 2H^+ + CO_3^{2-}$. The arrows indicate that the reactions can proceed in both directions: if CO_2 is increased, the reactions shift towards the right, producing H^+ ions, i.e. an increased acidity of the solution. A decrease in CO_2 makes the reactions go the opposite way. In a sense, these reactions behave as a series of communicating vessels.

[6] Brewer et al. [3].

ions (H^+); *pH* diminishes with the increasing concentration of hydrogen ions, i.e. with acidity. Acid solutions have a *pH* of 0 to 7; alkaline or basic solutions have a *pH* of 7 to 14; neutral solutions have *pH* of 7.0. Surface seawater is weakly alkaline with *pH*~8; such a condition favours calcium carbonate formation and allows organisms to develop a $CaCO_3$ exoskeleton. The situation changes in the deep ocean waters, where a high concentration of CO_2 lowers the *pH* to 7.6–7.5. Thus, at depths of approximately 3500–4500 m, known as the **Carbonate Compensation Depth** (CCD), seawater acidity is sufficiently high to destabilise and dissolve carbonate minerals. This phenomenon severely limits the deposition of carbonates in deep ocean environments.

Earth is the only planet of the solar system that contains liquid water on its surface. Its origin is contentious. Most believe water was incorporated during the planet's accretion from icy asteroids and other cosmic bodies some 4.56 billion years ago. Others suggest that water was delivered later on by comets; however, such a contribution seems minor on the basis of distinct isotopic compositions of water from comets and the Earth's oceans (Box 8.1).

Liquid surface water accumulated early in the Earth's history, likely some 4.4–4.2 billion years ago (Chap. 9). The early oceans were about 25% more voluminous than at present. Such a large reduction of the hydrosphere is a consequence of the slow transportation and sequestration of water in the deep layers of the Earth (Chap. 7), and the molecular destruction by photolysis which generates hydrogen that escapes to space. The temperature of the primitive hydrosphere was much higher than at present; *pH* was low, due to both the high concentration of CO_2 and intensive hydrothermal activity.[7] Free oxygen was virtually absent; therefore, many chemical elements such as nitrogen, iron, manganese, and uranium were present in the reduced state and had a different chemical behaviour than in the oxidising conditions of our era. For example, iron was prevailingly occurring as soluble divalent ion (ferrous iron, Fe^{2+}) and was abundantly dissolved in waters. The Great Oxidation Event led to the oxidation of divalent ferrous iron to insoluble trivalent ferric iron (Fe^{3+}), causing the accretion and precipitation of iron oxide-hydroxide minerals and their accumulation on the seafloor. Some of these deposits known as **Banded Iron Formation** (BIF) are still present today in some geologically ancient regions (e.g. Australia, Greenland, Brazil, Canada, India, etc.) and make up the largest reserves of iron ores in the world.

---

[7] Halevy and Bachan [4].

2.5   The Sedimentary Cycle: Air, Water and Earth at Work

The **sedimentary cycle** is a part of the larger **rock cycle**, briefly addressed in Box 7.1. It is conventionally divided into four main steps: rock weathering and erosion, transportation of the weathering products, deposition of the incoherent sediments and, finally, their compaction, lithification, and transformation into new rocks. Water is the leading agent of the sedimentary cycle, followed by wind. Other factors, such as glaciers, play a minor role.

2.5.1   Rocks Breakdown

Weathering is the process by which the solid rocks present at the Earth's surface are broken down and worn away by erosion. Weathering is accomplished by physical, chemical and biochemical processes, or by a combination of these.

Physical weathering is the mechanical breakdown of rocks that leads to the formation of separated mineral grains and rock debris. It is particularly efficacious on rocks with little or no soil cover and can be accomplished by repeated expansion and contraction prompted by temperature oscillations, or in the cold zones, by water infiltration in the fractures or pores, and expansion by freezing.

Chemical weathering is closely related to the action of water and atmospheric gases. Rainwater is slightly acid because of dissolved CO_2 and, therefore, can react with rocks and their constituent minerals, to form new compounds and transfer some chemical elements into solution (Box 2.1).

Biochemical weathering is accomplished by the biological activity of growing organisms, especially plants, lichens, algae and bacteria that attack rocks to extract nutrient elements.

Rates and intensity of weathering are more effective in hot or humid than in dry and/or cold regions. Acid rains and atmospheric pollution accelerate rock weathering, as evident from severe degradation of monuments and buildings in urban and industrial areas.

The alteration of silicate minerals through weathering has a major impact on the composition of the Earth's atmosphere and hydrosphere (Box 2.1). The main consequence is the long-term consumption of vast quantities of CO_2 that have been transferred from the atmosphere to sediments. Such a transfer is the cause of a drop in the concentrations of atmospheric CO_2 from thousands of parts per million by volume during the Precambrian to a few hundred parts per million today (Fig. 2.1). This descending trend has been irregular over geological time and associated with strong fluctuations in the global climate (Chap. 8).

2.5.2 Removal and Transportation of Weathering Products

The products of weathering include detrital minerals and rock debris or **clasts**, newly formed compounds such as clay minerals, oxide-hydroxide of iron, manganese, aluminium and other metals, and a large variety of chemical elements that dissolve as ions in the water (Box 2.1). A fraction of these alteration products remains in situ to form soils, but most are taken over by water and transferred to lakes, river flood plains, and the ocean (Fig. 2.3). The smaller the size of a material, the farther it can be transported away from its source by streams and rivers.

Larger clasts are left near to the source. Much of the fine- to medium-sized clastic fragments, such as gravels, sand and silt, tend to accumulate in the valleys, producing flat land (the **flood plain**). Some chemical elements, such as Na^+, Cl^-, Mg^{2+}, K^+, Ca^{2+}, Fe^{2+}, SO_4^{2-}, HCO_3^-, NO_3^-, and PO_4^{3-}, are transported in solution in the form of ions; other elements such as Si, Al, Fe^{3+}, Ti, Th, and U are poorly soluble and commonly precipitate as oxides-hydroxides during transport. This process drives a chemical selection that transfers elements to the sea in different amounts. As a result, seawater

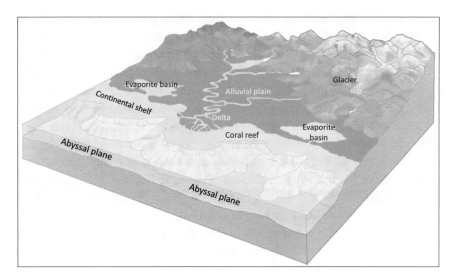

Fig. 2.3 Schematic pictorial view of the main sedimentary environments

becomes enriched in the most soluble elements, such as sodium and chlorine, and is depleted in others.

It is worth noting that some elements can have very different solubility, depending on their oxidation state. Notably, iron is soluble in its bivalent oxidation state but when is oxidised to trivalent ferric iron (Fe^{2+} → Fe^{3+}) becomes insoluble and precipitates as oxide/hydroxide. Therefore, the concentrations of iron in the oxidising environment of surface marine waters are minimal, with only a few micrograms per litre. Since iron is an essential nutrient for aquatic organisms, its scarcity dramatically limits the development of phytoplankton. Note that phytoplankton produces the largest part of atmospheric molecular oxygen and is the base of the marine food chain.

The wind is the second most important transportation agent. Dust that blows off from Earth's deserts is a primary carrier of chemical elements, including some nutrients such as iron and phosphorus that are transferred to oceans, to feed the phytoplankton. Wind-blown particles can accumulate on continents to form extensive "aeolian" deposits, such as the desert and littoral sand dunes, and the fine-grained ill consolidated **loess**, which covers with thick blankets vast areas of all continents.

2.5.3   Sediment Deposition

The sediments transported by water are deposited in both continental and marine environments (Fig. 2.3). Clastic fragments are deposited by mechanical settling at variable distances from their source, depending on grain size and the water flow energy. In particular environments, clastic deposition can also involve organogenic material, such as fragments of shells and corals.

Colloids settle down when small particles stick together to form larger aggregates (**flocculation**), for example, when river water reaches the sea. These aggregates are too heavy to be held in suspension, and thus sink and accumulate. Notably, many types of clay are products of colloidal settling.

The deposition of dissolved ionic species can only occur through chemical precipitation of salts in isolated sea basins or lakes, where evaporation exceeds the water influx from rivers and streams (**evaporite basins**). Typical evaporite minerals include Ca- and Mg-carbonate minerals, gypsum ($CaSO_4 \cdot H_2O$), halite ($NaCl$), sylvite (KCl) and borax ($Na_2B_4O_7 \cdot 10H_2O$).

Sediment deposition often occurs as discrete events that generate sequences of superimposed **layers** or **beds** or **strata**, each representing a distinct episode of sedimentation. Since sediments mostly settle down on flat surfaces such as lake or sea bottoms, the layers have an original horizontal aspect. The tilting, fracturing, and folding commonly observed in many sedimentary sequences cropping out at the Earth's surface are the consequence of tectonic stresses during uplift—a conclusion reached a few centuries ago by the Danish naturalist Niels Stensen (Nicolaus Steno in Latin; 1638–1686) and even earlier by Leonardo da Vinci (1452–1519).

2.5.4 Diagenesis: Back to Rocks

The final step of the sedimentary cycle is the consolidation of sediments and their transformation into lithified rocks. The process, known as **diagenesis**, involves the compaction of loose particles under the pressure of their weight, the expulsion of the interstitial water, the dissolution of some minerals and their partial re-precipitation in the interstitial spaces along the mineral grains. The newly-formed interstitial minerals make up the cement that holds the granules together, changing loose sediments into lithified rocks.

The clastic sedimentary rocks generated by accumulation and diagenesis of detrital material include **conglomerate**, **sandstone**, **siltstone**, and **claystone** made up of pebbles, sand, silt and clay, respectively. **Limestone** and **dolostone**—respectively formed of calcite and dolomite—are the main kinds of biogenic rocks. However, they can also result from the accumulation of carbonate clasts and chemical precipitation of carbonates. Dolostone is a sort of a 'secondary' rock that was initially made of calcite but had its composition changed during diagenesis through the addition of magnesium that partially replaced calcium. The accumulation of siliceous shells such as those of microscopic radiolarians or diatoms forms the **chert**.

Rock gypsum and **rock salt** are two well-known evaporite rocks that are found worldwide and represent economically essential sources of gypsum, halite, and other minerals. Other rocks formed by chemical precipitation of minerals include **travertine** and **geyserite,** respectively made up of carbonate and opaline silica accumulated around the geysers or thermal springs, and **stalactite** and **stalagmite** that form on the insides of karst caves through the deposition of minerals especially calcite.

2.6 The Remains of Mountains

The terrestrial landscape is the most remarkable outcome of the sedimentary process. Erosion flattens mountainous relief and excavates valleys; surface water transports sediments to the lowlands and tends to fill them. The end result should be a perfectly spherical Earth (which would be the minimum surface-energy landform) completely enveloped by water, as early envisaged by Leonardo da Vinci in the fifteenth century. However, such an end-state hasn't been attained because of the large-scale movements of the Earth's crust (**tectonic movements**) and volcanism that throughout Earth history have continually uplifted mountain ranges, opened new ocean basins, and constructed volcanic cones.

Extensive erosion is the final fate of all kinds of topographic relief. Many Precambrian mountainous terrains, once appearing like today's Alps and Himalaya, are now reduced to flat or weakly undulating landscapes known as **peneplains**. Such features are found on all continents, from the so-called 'Canadian Shield' of east-central North America, to various regions of Africa, Eurasia, Australia and Antarctica. The outcropping rocks on peneplains represent core zones of ancient mountain belts, finally exposed to the surface after many hundred millions years of erosion. These old cores of mountain ranges consist of a complicated assortment of different rock types. The occurrence of multiple folding and different generations of fractures and magmatic bodies testify to the long and complex sequences of tectonic processes, magmatism, metamorphism, deformation, erosion, and sedimentation suffered by these ancient mountain belts millions or billions of years ago (Chap. 6).

Landform geometries are endless. Their variability results from numerous factors such as the mechanical strength of rocks, the kind of tectonic structures present in an area, and climate. Even a rough analysis of these issues is beyond the scope of this book, but there are many geomorphology books that provide excellent information on the subject.

2.7   The Fate of the Organic Matter

Both inorganic (shells, coral exoskeletons) and organic soft components are involved in the biogenic sedimentation. Accumulation of the inorganic remains of marine organisms forms limestone and chert, whereas the accumulation and transformation of organic soft components may lead to the formation of hydrocarbons and coal.

Organic components of aquatic organisms mostly accumulate on the shallow seafloor of continental shelves, but they can also deposit in continental and transitional basins such as lakes, wetlands and deltas. In these environments, water energy is moderate and there is a wide range of organisms—from marine microplankton, urchins, molluscs, and crustaceans to a variety of vegetation. If sedimentation and burial are quick, and sediments are fine-grained, organic material is rapidly isolated from the external environment, escaping oxidation and slowly modifying to hydrocarbons. These transformations take months or years to form marsh gas and hundreds of thousands to millions of years to form oil.

Large **hydrocarbon** deposits derive from marine microplankton. The transformations that occur after burial generate methane and some organic compounds called **kerogens**. At temperatures between about 70–150 °C, oil is formed; gaseous hydrocarbons such as methane and butane then form at higher temperatures. Once formed, oil and gas can slowly migrate from the source rocks to the surface or be trapped within some rock formations to accumulate as hydrocarbon deposits (Fig. 2.4). To form a good physical trap, a rock must have high permeability and be covered by thick impermeable layers, such as clay or shale. Such conditions favour hydrocarbon accumulation and prevent its dispersion to the surface.

Hydrocarbons can also remain in the source rocks if these are impermeable, i.e. do not have interconnected porosity. Under these conditions, oil and gas remain trapped within the rocks, such as has happened in the so-called **shale oil** and **shale gas** fields being exploited around the world today.

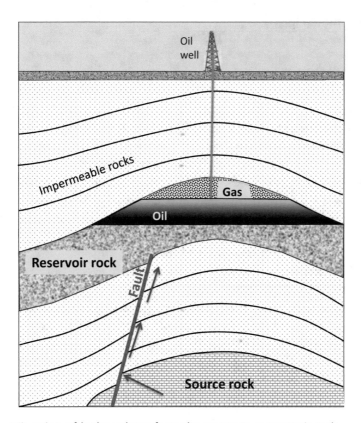

Fig. 2.4 Migration of hydrocarbons from the source to a reservoir rock

Fossil coal forms on continents when dead plant matter becomes submerged in swamps and is subsequently subjected to burial. An increase in pressure and temperature slowly leads to water loss and an increase in carbon content. The most valuable type of coal is **anthracite,** which contains 75% to 90% carbon and is generated at temperatures between about 200–300 °C. Coal varieties of lower grade, such as **lignite** and **peat** respectively containing up to 70% and less than 60% carbon, are formed at lower temperature.

To conclude, hydrocarbons and coal are made up of chemical elements, mostly carbon, extracted from the atmosphere and water by terrestrial and marine organisms during their lifetime and buried in the subsurface for geological times. In other words, **fossil fuel** is old carbon that was withdrawn from the fluid Earth and thereafter slowly transferred to become part of the solid Earth. The total amount of carbon in fossil fuel and other organic sedimentary material is estimated to be about 20 million gigatons.[8] Besides, massive amounts of inorganic carbon (about 70 million billion tons) is stored in limestone (Box 2.2). All of this carbon has been subtracted to the atmosphere by sedimentary processes, thus explaining the drop in CO_2 during geological time (Fig. 2.1).

2.8   Summary

Rock weathering, the transportation of its decomposition products, and the subsequent deposition of sediments in valleys, lakes, and seas are the main geological processes occurring at Earth's surface. They make up what geologists term the 'sedimentary cycle'. Water is the main agent of the cycle, whose action shapes the Earth's landscape, forms the soils, and continually modifies the composition of the hydrosphere and atmosphere. Wind also plays a key role in the sedimentary process, especially in the accumulation of widespread loess deposits and continental and littoral sand.

Weathering—the decomposition of rocks and minerals—is performed by physical and chemical processes. Water is the main agent, but the activity of living organisms also plays an important role.

The products of rock weathering include detrital fragments (boulders, cobbles, pebbles, and sand), silt and colloidal-size particles (clays, oxides and hydroxides of various elements), and a large number of dissolved elements present as ions (Na^+, Ca^{2+}, Fe^{2+}, HCO_3^-, Cl^-, etc.). They are transported

---

[8] One gigaton (Gt) is one billion tons.

for variable distances and accumulated in the lowlands or delivered to the ocean.

Oceans are the primary place of sediment deposition. The chemical composition of ocean water is essentially a consequence of the addition of ionic species that have been extracted from the rocks and soils and transported to the sea by rivers. Some of these elements are nutrient for living organisms and, therefore, provide the base material for marine life. Additional factors that determine seawater compositions include direct exchange with the atmosphere and ocean crust, submarine volcanic and hydrothermal activity, addition of windblown dust, and biological processes.

The reactions between atmospheric gases and rocks consume large quantities of carbon dioxide (CO_2), which is transported to the sea as bicarbonate ions (HCO_3^-) and finally deposited onto the seafloor by chemical or biochemical activity. These processes have been active throughout most of Earth history, i.e. over billions of years, and have resulted in a net reduction of atmospheric CO_2 from thousands of parts per million by volume during the Archaean aeon to about 260–270 parts per million by volume in pre-industrial times.

Living organisms play a key role in the sedimentary cycle. They facilitate the alteration of rocks and the formation of soils. The deposition and burial of their hard shells in marine environments generate limestone and chert, whereas accumulation and rapid burial of soft parts can lead to the formation of hydrocarbon. Deep burial of plants in marshy continental areas and heating at depth give coal. Together, these processes represent the primary mechanism for the subtraction of chemical components, especially carbon, to the fluid Earth and its long-term storage inside the lithosphere.

2.9   Box 2.1—Water–Air-Earth Interaction: Some Basic Geochemistry

To better understand chemical weathering and its effects, it is necessary to consider simple chemical reactions that involve atmospheric gases, H_2O, and the most common rock-forming minerals. Some examples are as follows:

$$Mg_2SiO_4 \, [olivine] + 4CO_2 + 4H_2O \rightarrow 2Mg^{2+} + 4HCO_3^- + H_4SiO_4;$$
$$(2.1)$$

$$2(Na, K)AlSi_3O_8[alkali\ feldspar]$$
$$+ 2CO_2 + 11H_2O \rightarrow Al_2Si_2O_5(OH)_4[kaolinite]$$
$$+ 2(Na^+,\ K^+) + 2HCO_3^- + 4H_4SiO_4; \tag{2.2}$$

$$FeSiO_3\ [Fe - pyroxene] + 2CO_2 + 3H_2O \rightarrow Fe^{2+} + 2HCO_3^-$$
$$+ H_4SiO_4; \tag{2.3}$$

$$CaSiO_3\ [Ca - pyroxene] + CO_2 \rightarrow CaCO_3\ [calcite]$$
$$+ SiO_2\ [silica] \tag{2.4}$$

$$Ca_5(PO_4)_3(OH)\ [apatite] + 4CO_2 + 4H_2O \rightarrow 5Ca^{2+} + 3HPO_4^{2-}$$
$$+ 4HCO_3^- + H_2O; \tag{2.5}$$

$$CaCO_3\ [calcite] + CO_2 + H_2O \rightarrow Ca^{2+} + 2HCO_3^-; \tag{2.6}$$

$$4Fe^{2+} + O_2 + 6H_2O \rightarrow 4FeOOH\ [goethite] + 8H^+ \tag{2.7}$$

The reactants are on the left side of the arrows, whereas reaction products are on the right. The implications of these reactions for the external terrestrial environment are of paramount interest and can be easily understood by anyone with little knowledge of chemistry:

a—Almost all reactions involve the consumption of carbon dioxide (CO_2), which is always present among the reagents. Rainwater scrubs CO_2 out of the atmosphere, producing bicarbonate ions (HCO_3^-), some of which are consumed in the weathering of rocks and the remainder is transported in solution to the sea and then used by organisms to build up carbonate shells and exoskeletons. The accumulation on the seafloor of such carbonate gives rise to organogenic limestone, which retains CO_2 for tens or hundreds of million years. Carbonate rocks, therefore, represent an almost permanent major sink for CO_2 (Box 2.2).

b—Metal ions (Na^+, Mg^{2+}, Ca^{2+}, etc.) are almost always formed as reaction products that are dissolved in the water and transported to the sea by rivers. Oceans' salinity is substantially determined by the accumulation of the ions removed from rocks and soils. Some species, such as potassium, phosphorus, nitrogen, sulphur, silicon and iron produced from the weathering of alkali feldspars, apatite, pyroxenes, and sulphates are nutrients for marine plankton. Therefore, rocks and minerals are the primary sources of the nutrient elements that feed marine life.

c—Some weathering reactions produce solid substances, such as kaolinite, calcite, and metal oxides-hydroxides, that appear on the right side of Eqs. (2.2), (2.4), and (2.7). These are left in place to build up the soil or are transported in suspension by water as colloidal particles. The muddy colour of streams and rivers, especially during heavy rain periods, is mainly due to colloidal material.

d—Reaction (2.6) refers to the dissolution of calcite, a mineral almost insoluble in pure water. However, the addition of atmospheric CO_2 causes a slight acidification of rainwater, which can slowly dissolve calcium carbonate. An essential aspect of this reaction is that it does not produce any solid residues, which explains why karst terrains, consisting of carbonate rocks, have little or no soil and are relatively bare of vegetation.

e—Reaction (2.7) represents the oxidation of divalent ferrous iron, a fundamental constituent of many minerals; the process produces a change in the colour of the rocks that become red-yellow or rusty in colour on the external surfaces in contact with the atmosphere. Geologists use the hammer to break the alteration skin and observe the "fresh" material inside.

For a complete picture, the following reaction should be considered:

$$Ca^{2+} + 2HCO_3^- + H_2O \leftrightarrow CaCO_3 + CO_2 + 2H_2O.$$
$$calcite \tag{2.8}$$

This reaction designates the precipitation of calcite, the opposite of reaction (2.6). The mineral precipitates when the solutions become super-saturated in calcium carbonate, for example, by loss of dissolved CO_2. The phenomenon is common at thermal springs or in the karst caves, in which travertine, flowstones, stalactites, and stalagmites are formed.

Comparing reactions (2.6) and (2.8) makes it clear that the amount of CO_2 consumed by the dissolution of calcium carbonate is identical to that released by its crystallisation (i.e. one CO_2 molecule). It follows that the alteration of calcite does not lead, by itself, to a net loss of atmospheric CO_2 in the long term, since the amount consumed during its dissolution is compensated by that released during precipitation. By contrast, silicates are net consumers of CO_2.

Most of the reactions described above have profound climatic implications that all stem from the well-established power of CO_2 to trap heat close to the surface of the Earth, inducing the warming of the lower atmosphere. This is referred to as the **greenhouse effect**, a topic that will be discussed further in this book.

Alteration of silicate minerals is the main mechanism to consume CO_2 and permanently remove it from the atmosphere. By contrast, volcanic outgassing works the opposite way, adding CO_2 to the fluid Earth. If silicate weathering predominates, the concentration of CO_2 in the atmosphere decreases and the global climate cools down.[9] By contrast, if volcanic degassing predominates, CO_2 concentrations increase, the greenhouse effect intensifies, and the global climate becomes warmer. Warm climate, in turn, accelerates mineral alteration, which consumes higher amounts of atmospheric CO_2, resulting in a cooling of climatic conditions down. In conclusion, the balance among these factors acts as a sort of natural thermostat that keeps the Earth's climate stable.

2.10   Box 2.2—Limestone, a Most Meritorious Rock

Chemical and biogenic rocks represent the pool of chemical elements extracted from the lithosphere and atmosphere and transferred to sediments. Carbon is the element that is most heavily involved in such a relocation.

Carbon occurs in many different natural compounds, both inorganic (e.g. calcite, dolomite) and organic (cellulose, proteins, etc.). In the atmosphere, it is mainly present as carbon dioxide CO_2; during the sedimentary cycle, it passes to the hydrosphere as bicarbonate ion (HCO_3^-) through the alteration of silicates, or directly dissolves as CO_2 in the ocean water. Here, carbon is involved in the vital cycle of aquatic organisms, whose remains form hydrocarbons and organogenic rocks, especially limestone.

Carbonate rocks on Earth contain around 70 million billion tons (or petatons) of carbon. If limestone did not exist, these enormous quantities would be retained entirely as CO_2 in the hydrosphere-atmosphere systems. The consequences for the external environment of our planet would be dramatic, given the well-established greenhouse effect of this gas.

Therefore, limestone is indeed a high meritorious rock: thanks to it, the Earth is a hospitable planet, rather than an immense greenhouse with surface temperatures of hundreds of degrees, such as is the case on our sister planet, Venus.

Let's take a closer look at this providential rock. Calcite ($CaCO_3$) is its dominant mineral. Dolomite [$(MgCa(CO_3)_2)$] is the second most abundant mineral and is the main constituent of dolostone—the rock that builds up the splendid scenery of the Dolomites in the Italian Southern Alps. The chemical formulae of calcite and dolomite can also be expressed in oxides [$CaOCO_2$

---

[9] For instance, this happens when large masses of fresh rocks are exposed at the surface during the uplift of great mountain ranges, such as the Alps and Himalayas.

and $CaOMgO(CO_2)_2$], a format that better highlights that carbon dioxide (CO_2) is the main component of these minerals.

Limestone generally has a whitish colour, but there are black, brown, greenish, or varicoloured varieties due to the presence of impurities such as clay minerals, iron oxides-hydroxides, or organic material. It is often a well-cemented, coherent rock and an excellent building material; its hardness is moderate and the rock can be easily scratched by metal objects, for example by the geologist's hammer. When in contact with an acid, such as diluted hydrochloric acid (HCl), limestone emits an intense effervescence due to the release of CO_2 from the reaction with calcite. Dolostone shows the same phenomenon only if reduced to fine powder. This is a demonstration that CO_2 is indeed a main component of carbonate minerals. If subjected to high temperature and pressure inside the Earth (i.e. metamorphism), limestone turns into marble; this retains the original mineralogical compositions of its parent rocks but develops a more compact structure.

Seafloor is the primary deposition environment of limestone, especially at medium to low depths of continental shelves. The deposition rate changes significantly from hot to cold settings; therefore, the abundance of carbonates in the sedimentary sequences provides important information on paleoclimate. Limestone retains a memory of its original deposition environment; particularly, the relative abundances of oxygen and carbon isotopes vary according to the chemical-physical conditions of formation; this allows determining temperatures and concentrations of atmospheric CO_2 in the past (Box 8.1). Finally, limestone is home to countless fossil remains, to which we owe much of what we know about the evolution of life on Earth.

Numerous monuments and statues of our cities have been built with limestone and marble. There is probably no monument dedicated to these rocks in recognition of their unique role in the evolution of the Earth's environment. While we wait for humanity to remedy its negligence, it is wise to pay tribute to these beneficial rocks by showing a picture of the Capo Caccia, in the province of Alghero, Sardinia (Italy), a magnificent example of Mesozoic carbonate rock sequence (Fig. 2.5).

Fig. 2.5 Mesozoic limestone cliff of Capo Caccia (Province of Alghero, Sardinia). Image courtesy of Pianeta Alghero

References

1. Royer DL, Berner RA., Montanez IP, Tabor NJ, Beerling DJ (2004) CO_2 as a primary driver of Phanerozoic climate change. Geol Soc Am Today 14:4–10
2. Krauskopf KB, Bird DK (1995) Introduction to geochemistry. McGraw-Hill, p 647
3. Brewer PG, Friederich G, Peltzer ET, Orr FM Jr. (1999) Direct experiments on the ocean disposal of fossil fuel CO_2. Science 284:943–945
4. Halevy I, Bachan A (2017) The geologic history of seawater pH. Science 355:1069–1071

3

Fire—How Magmatism Shaped the Earth

[...] and where there are active volcanos, there must exist, at some unknown depth below, enormous masses of matter intensively heated and, in many instances, in a constant state of fusion. We have then to inquire, whence is this heat derived?
Charles Lyell, Principles of Geology (1850)

3.1 Introduction

The Earth is a layered planet, from the inner core to the outer extremes of the atmosphere. Such a stratified structure is the effect of the separation of chemical elements and compounds that were at the beginning of Earth history—about 4.5 billion years (Ga) ago—mixed in a single hot liquid mass, but successively separated into distinct superimposed spherical shells or layers. Simultaneously, immense quantities of heat were lost from inside the planet to space, finally resulting in an ordered, relatively cool and stratified planetary body. Magmatism has been the prime driving force of these colossal transformations.

The complexity of magmatic processes is challenging, but only the most interesting and easily accessible aspects will be discussed in this chapter. In particular, the focus will be on the role of magmas in constructing a layered Earth and its impact on our external environment; some basic principles of physical volcanology and the global distribution of magmatism will also be addressed briefly. For a better understanding of the subject matter, a short overview of magma origin and evolution is given in Box 3.1. Box 3.2 and Box 3.3 describe heat transfer mechanisms operating from the inside to the surface of the Earth, and the origin of ore deposits—two topics closely related to magmatism.

3.2   Magmas: What They Are, How They Form

Magmas are high-temperature natural liquids (T ~ 800–1200 °C) that originate inside the Earth by partial melting or **anatexis** (ἀνάτηξις, anátexis = liquefaction) of rocks. Therefore, they can be simply defined as molten rocks.

Magmas originate inside the upper mantle or the middle-lower continental crust and, once formed, move toward the surface. Most of the times, magmas pond inside the crust or at the crust-mantle boundary forming large accumulations referred to as **magma chambers**. Here magmas cool down and partially solidify before eventually being erupted on the surface. The largest fraction of magmas in nature completely solidifies at depth, and only about 20% of the total reaches the surface.

The crust and mantle are made of solid rocks. Nonetheless, they can melt locally, in particular physical–chemical conditions, e.g. when there is a decline in the ambient pressure, a rise in temperature, or the introduction of water and other fluid substances into the hot rocks (Box 3.1).

The generation of magma is a sort of geological anomaly that takes place at restricted places inside the Earth. Volcanoes reveal that something unusual is taking place at depth. Similarly, the occurrence of volcanism in the geological record tells us much about the past history of our planet.

Almost all Earth magmas have a silicate composition dominated by silicon and oxygen, with variable quantities of nearly all chemical elements, including sodium, magnesium, aluminium, phosphorus, potassium, calcium, titanium, manganese, and iron. A high number of very diluted components (**trace elements**) such as nickel, rubidium, strontium, uranium, thorium and many others, are present in magmas at parts per million (*ppm* = grams of element per ton of rock) to parts per billion (*ppb* = milligrams of element per ton of rock) levels.

Magmas also contain variable quantities of dissolved gas. Water is the most abundant, followed by CO_2. Other gases such as H_2S, SO_2, CO and HF occur in smaller amounts, although a pungent odour makes sulphur compounds the only ones to be perceived in active volcanic areas. Gases are incorporated into the magmas during formation in the source zones, but significant quantities can be collected from wall rocks during the ascent.

The behaviour of H_2O and CO_2 in magmas is intriguing and worth emphasising. Water dissolves easily both at moderate and high pressures; by contrast, CO_2 is only soluble at very high pressures of 2–3 gigapascal (GPa). Therefore, when H_2O- and CO_2-rich magmas rise from the Earth's interior to the surface, they release first carbon dioxide at depth, but carry dissolved water up to the surface. In other words, carbon dioxide is almost completely lost deep inside the Earth, whereas water is released in the atmosphere during volcanic eruptions. Such contrasting behaviour has paramount consequences for the Earth's external environment: if the reverse happened, there would be a lot of CO_2 and very little or no water on the surface of the Earth!

The types of magmas occurring on Earth are many and change from one place to the other, or also sometimes within individual volcanoes. **Basalt** is the most common one, followed by **granite** (or its effusive compositional equivalent **rhyolite**) and **andesite**; their compositions are also referred to as **basic, acid,** and **intermediate**,[1] respectively.

Basaltic magmas have the highest temperatures (about 1100–1200 °C) and density (2.7–2.8 g/cm^3), are very fluid, and contain moderate quantities of gaseous components. Granite magmas have lower temperature (around 800–900 °C) and density, are highly viscous, and generally contain higher amounts of gas than basalts. Andesitic magmas show intermediate chemical-physical characteristics between granite and basalt. Variable properties of magmas are important because they are the cause of contrasting modalities of eruptions.

The cooling and solidification of magmas produce **igneous** rocks, one of the three fundamental rock families existing on Earth (Box 1.1). Those formed by magma solidification at depth are called **intrusive**, whereas rocks derived from magma erupted onto the surface are named **volcanic** or **effusive**. Igneous rocks are the primary source of information about the physical characteristics and composition of magmas.

3.3   Magmatism and the Structure of the Earth

Partial melting of rocks inside the Earth, separation of the melts from the solid residue, and rise towards the surface are the three main steps of magmatic processes. Figure 3.1 explains how it all happens, and the implications for the planet Earth.

---

[1] These are archaic but still widely used terms that originated in the mid-nineteenth century when naturalists thought that the high quantity of silicon in the granite and rhyolite was present as silicic acid H_4SiO_4, whereas Ca and Mg were present as bases (metal hydroxides). We now know that this is not true, yet the terminology has remained well-entrenched in the geological literature.

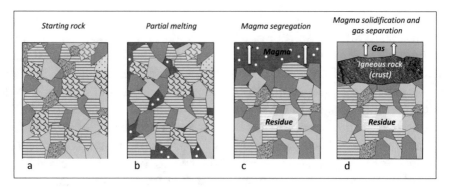

Fig. 3.1 Main stages of magma formation, segregation and ascent. **a** Starting rock inside the Earth made of various grains of different minerals indicated with distinct patterns and colours; **b** Partial melting of some minerals and formation of a gas-rich magma (red colour with white dots); **c** Segregation of the gas-rich magmatic liquid from the residual rock and its upward migration; **d** Solidification of magma at the surface to form igneous rocks and release gas into the surroundings

A mantle or crustal rock made up of various types of minerals is the starting material (Fig. 3.1a). When the conditions are right—i.e. there is an increase in temperature, pressure drop or addition of water—the rock undergoes melting; however, only a few minerals melt, while most remain in the solid state (Fig. 3.1b). The magmatic liquid has a lower density than the coexisting solid; therefore, it separates from the source by floating and migrates upwards (Fig. 3.1c), until it reaches the shallow crust or the Earth's surface where it solidifies and releases its gas load (Fig. 3.1d).

Any type of magma has a very different chemical composition than the parent rock and the residual solid, and is particularly enriched in gas, silicon, potassium, phosphorous and many other elements. These elements are referred to as **incompatible** or **magmatophile** (literally 'magma-loving'), because they are too big or highly charged to be hosted in the crystal lattice of the residual minerals and prefer to go into the magma (Box 3.1).

The crux of this sequence of processes is that partial melting of a single material (the starting rock) leads to the formation and separation of three physically and compositionally distinct systems: the solid residue, the magmatic rock, and the gas phase; *ex uno plures*, one could say by reversing a well-known motto.

The significance of the mechanism described above for the evolution of the Earth system is straightforward. The peridotite making up the primordial mantle is the starting rock. Partial melting of peridotite generates basaltic liquids with a much lower density than the parent rock ($\sim$2.8 g/cm^3 *versus* $\sim$3.3 g/cm^3). Basalt magmas separate by density contrast and rise towards the

Earth's surface, leaving in the mantle a solid residue depleted in some chemical components. Cooling of magmas on the Earth's surface generates solid rocks (crust) and releases the gaseous components. The replication of this process again and again during the history of our planet ultimately resulted in the Earth's stratified structure, consisting of three compositionally distinct layers: the mantle, crust, and the fluid Earth (atmosphere-hydrosphere).

A remarkable effect of these differentiation processes is that melting refines the mantle rocks by collecting magmatophile elements and taking them up to the surface to form the crust. As a result, the crust is enriched in silica, potassium, phosphorous and other elements, some of which are nutrients to living organisms. When crustal rocks are weathered, they release their constituent elements into the sea, thus fertilizing life in the oceans. The vegetation on land also extract nutrients from the rocks. Therefore, magmatism ensures a continuous supply of nutrient elements from the interior to the surface of the Earth, representing a major promoter of the development and evolution of life.

This is the essence of the contribution of magmatism to the shaping of the Earth and its external environment. Such a process took a long time to complete: a complex but interesting story that is worth examining in some detail.

3.3.1 From Chaos to Order

The Earth began to form by the aggregation of cosmic particles about 4,56 million years (Ma) ago. A particular class of meteorites, called **chondrite**, provided most of the materials that built up the planet (Chap. 9).

Meteorite impacts, compaction of the aggregated material, and decaying of the radioactive elements generated large amounts of heat that melted the planet, generating a molten sphere continuously mixed by convection movements (Fig. 3.2a). The bulk composition of the proto-Earth was homogeneous, matching that of the average chondrite.

Successively, the liquid iron-nickel separated from silicates and sank into the centre of the Earth to form the core, leaving lighter silicates to migrate upwards and form a magma ocean (Fig. 3.2b). The segregation of iron and its sinking towards the centre can be considered the first global magmatic differentiation event of our planet.

The magma ocean condition did not last for long, as the loss of heat to space caused the formation of a thin, solid crust, which was followed by the slow crystallisation of the Earth's interior. Mantle solidification expelled large quantities of gaseous substances that could not be hosted in the solid rocks.

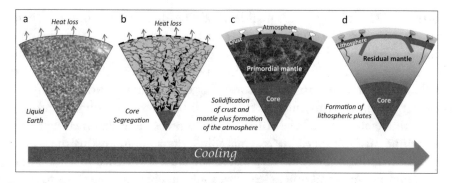

Fig. 3.2 Main stages of Earth evolution. (a) The primordial Earth was a hot liquid body; (b) At an early stage, liquid blobs of metallic iron-nickel separated from less dense silicates and sank towards Earth's centre, forming the core; (c) Heat loss progressively led to solid mantle and crust, and a gaseous atmosphere; (d) Further cooling and continuous magmatic activity modified the primitive composition of mantle, crust, and fluid Earth. At some stage, a thick and rigid lithosphere was formed, which broke up into mobile plates, starting with modern-style plate tectonics (Chap. 6)

These gases concentrated in a primitive atmosphere, completing the second event of large-scale magmatic differentiation.

Subsequently, Earth evolution was dominated by melting events at different spots inside the primitive mantle and the primordial crust. Basaltic magmas were formed in the mantle and ascended towards the surface, increasing the mass of the crust and adding gaseous components to the atmosphere. The mantle became progressively depleted in gas and magmatophile elements, continuously modifying its initial composition (Fig. 3.2c). Partial melting also affected the crust, leading to the formation of granite magmas whose solidification produced the first small continental masses.

An essential step in the evolution of the planet occurred around 3.0 billion years (Ga) ago, when cooling generated a thick and rigid lithosphere that broke into several mobile blocks or **plates**, starting the modern-style plate tectonics (Fig. 3.2d). Lithospheric mobility affected very radically the entire silicate Earth, causing the formation of mountain belts, opening of oceans, growth of continents, and crustal reworking by sedimentation, metamorphism and partial melting. Magmatic compositions changed radically, and significant quantities of andesite magmas started to erupt, along with basalts.

The transfer of magmatic matter from the inside to the surface of the Earth has been a continuous and unidirectional process. Its persistence over the last billion years should have led to severe depletion of gas and magmatophile elements in the mantle, leaving a geochemically sterile rock, incapable of forming new magmas. However, another process has been operating in the opposite direction to reintroduce fluid substances and magmatophile elements back into the mantle. Such a process is called **lithospheric subduction**, which will be better explained in Chap. 6.

Subduction is a tectonic process peculiar to our planet and a key factor in its evolution. It consists of the sinking of the crust that returns the incompatible elements and gases (Si, K, Rb, Th, U, rare earth elements, H_2O, CO_2, etc.) back into the mantle, partially reintegrating what was extracted during the formation and separation of magmas. Subduction is a complementary process to magmatism. Together, they act in concert to establish a nearly balanced steady-state condition for the Earth system, making the amount of element input for each sphere similar to the amount of its output.

The present-day Earth dynamics can, therefore, be considered a competition between large-scale geological processes operating in opposite directions—(1) magmatism that facilitates selective migration of chemical elements, generating a compositionally ordered stratification and (2) subduction that transports this material back to depths, favouring mixing and the increase of entropy.[2]

The bulk composition of the Earth has not changed during its long history, except for the limited loss in the space of light elements, such as hydrogen and helium, and the daily arrival of cosmic dust and meteorites[3]. The early chaotic chondritic Earth has turned into an ordered and layered chondritic Earth. This transition from disorder to order is in apparent contradiction with both common sense and the second law of thermodynamics. This principle states that all systems tend toward a state of disorder and a chaotic system cannot spontaneously transform itself into an ordered one. More simply stated, it's easy to make an omelette from an egg, but not an egg from an omelette. However, the transition from disorder to order is possible through the addition of energy. In the case of Earth, the energy required to construct an ordered and stratified planet from a chaotic molten sphere comes from gravity and heat loss.

---

[2] Anderson [1].

[3] The mass of cosmic material that falls on Earth is estimated to be between 10,000 and 100,000 tons per year (10^7–10^8 kg). This quantity is tiny compared to the mass of the Earth (5.98 x10^{24} kg).

3.4   Volcanism

Volcanism is the external manifestation of magmatism. Volcanoes are formed along deep crustal faults that allow magmas to rise from source to the surface. Faults provide the pathway, but magma ascent is caused by buoyancy forces.

Volcanic eruptions range from effusive to strongly explosive. The former consists of the gentle discharge of lavas, whereas the latter takes place by violent outburst and ejection of magma and rock fragments (**pyroclasts**). Gaseous emissions are copious during volcanic eruptions, but they continue, more quietly, during periods of dormancy.

The eruptive style of volcanoes depends on magma composition, viscosity, and gas content. Because of their low viscosity and relatively low gas contents, basalts tend to erupt effusively, with little explosive activity. However, explosivity can increase dramatically during **hydrovolcanic** or **phreatomagmatic eruptions**, when basalt magmas accidentally come into contact with surface waters in volcanic conduits or at the eruptive vents. Intermediate and acid magmas are more viscous, have high gas contents, and therefore erupt much more violently than basalts.

3.4.1   Types of Volcanic Eruptions

Based on the explosive energy and the amount of erupted material, volcanologists distinguish different types of eruptions.

Hawaiian eruptions are typical of those occurring at Kilauea, the most active volcano on the Island of Hawaii. They are characterised by the quiet release of basaltic lava accompanied by modest explosive phenomena, such as the ejection of molten magma chunks and lava fountains. Some Hawaiian eruptions are gigantic, continuing for years and discharging huge volumes of lava and gas. Hawaiian eruptions taking place along elongated fractures are referred to as **Islandic**. Massive outpourings by islandic eruptions can fill the valleys, flatten the topography, and build up lava plateaux.

Strombolian eruptions get their name from the Island of Stromboli in Southern Italy, where semi-molten fragments of moderately viscous lava (**scoria**) are rhythmically ejected from the vent, solidify as they fly through the air, and accumulate around the crater after relatively short ballistic trajectories.

Vulcanian eruptions are typical of gas-rich andesitic and rhyolitic magmas. They are characterised by a series of fairly violent explosions that form 'cauliflower'-shaped eruptive clouds of gas and pyroclastic material (ashes, lapilli and blocks). The archetypal eruption of this type took place at Vulcano,

Southern Italy, in 1888–1890 and was described in detail by the Italian volcanologists Giuseppe Mercalli (1850–1950) and Orazio Silvestri (1835–1890), two fathers of Volcanology.

Surtseyan eruption is a violent hydrovolcanic phenomenon that occurs when magma interacts with external water. It is named after the volcano Surtsey, which erupted off the southern coast of Iceland in 1963–1967. Surtseyan activity is characterised by repeated strong explosions that expel jets of pyroclastic material that exhibit an overall 'cock's tail' or 'cypressoid' shape.

Plinian eruptions are the most violent and massive. They occur when large bodies of magma pond for a long time in the magma chambers below the surface, cooling and crystallising extensively. The residual liquid becomes rich in gas and light enough to open up cracks in the roof rocks and quickly rise to the surface. The eruption rapidly forms a massive column of ashes, pumice, and gas that can rise to more than 10 km in the atmosphere, expanding at its top and taking a 'pine tree' shape. The most famous Plinian eruption is that of Vesuvius in 79 A.D., which destroyed Pompeii, Herculaneum, and the surrounding areas. This eruption was accurately described by Pliny the Younger in two letters to the historian Tacitus. Some exceedingly large Plinian eruptions, referred to as **Ultraplinian**, can release tens to hundreds or thousands of cubic kilometres of ash and pumice. Thankfully, these eruptions are rare and only occur at some giant volcanoes (**supervolcanoes**), such as Yellowstone (Western USA), Tambora (Indonesia), and Campi Flegrei (Southern Italy).

3.4.2 Volcanism and the Terrestrial Environment

Volcanism has significant effects on the external environment. Long-term degassing has gradually modified the composition of the atmosphere and hydrosphere. However, gigantic volcanic eruptions have the potential to force the global climate, by emitting huge amounts of fine powders, CO_2 and sulfur gases.

Ashes ejected by Plinian and Ultraplinian eruptions remain in the air for months or years and can amplify the Earth's albedo (the proportion of incident sunlight that is radiated back into space), cooling the troposphere. Discharge of massive amounts of potentially dangerous gas species also affects the environment. SO_2, HCl and HF all contribute to the formation of acid rain and the reduction of stratospheric ozone (O_3). A particular impact on climate is related to the aerosol of sulphuric acid, which is formed in the stratosphere by the reaction between the volcanic sulphur oxides and water; such aerosol absorbs and reflect incoming solar radiation, hampering its arrival to the troposphere and lowering global temperature for months or even years. Equatorial eruptions are able to change the climate globally, whereas the effects of mid- to high-latitude events are generally restricted to their hemisphere of occurrence. The sulphate aerosol created by the eruption of the Tambora volcano (Indonesia) in 1815 caused the average surface temperature to drop by more than 1 °C causing severe damage to agriculture, so much so that people called 1816 the *Year of Hunger* or *Year Without a Summer*. The Hawaiian-type eruption of the Laki volcano (Iceland) in 1783 contained vast quantities of poisonous gases that spread across Europe and created famine preceding the French revolution. Multiple eruptions closely spaced in time can also have global climatic effects on a centennial time scale. Many scientists believe that such a sequence of events forced climate between the fourteenth and eighteenth centuries, triggering the so-called Little Ice Age.[4]

Finally, volcanoes are strong emitters of carbon dioxide (CO_2)—one of the natural gases producing the 'greenhouse effect'. The high quantities of CO_2 released into the atmosphere by widespread volcanic activity during the Precambrian had a heating effect that prevented the planet from freezing completely when solar radiation was about 30% lower than today. The numerous periods of glaciation that affected the Earth, both in the Precambrian and Phanerozoic aeons, have been accompanied, and sometimes caused by a decrease in atmospheric CO_2. Conversely, the accumulation of CO_2 from volcanic eruptions may have contributed to the end of glacial episodes, inverting the trend of temperatures and leading to a return to more comfortable climatic conditions (Chap. 8).

3.4.3   Large Igneous Provinces and Mass Extinctions

At several times throughout Earth history, gigantic volcanic events occurred over a geologically short time span of a few hundred thousand years, producing dramatic effects on the Earth's external environment. These eruptions produced an exceptionally high volume of basaltic (more rarely rhyolitic) magmas that were emplaced as thick sequences of lava flows over vast regions, both on the seafloor and the continents. These areas are known as **Large Igneous Provinces (LIP)**.[5]

The total volume of volcanic products in a single LIP can be millions of cubic kilometres that extend over hundreds of thousands to millions of square

---

[4] Briffa et al. [2].

[5] Saunders [3].

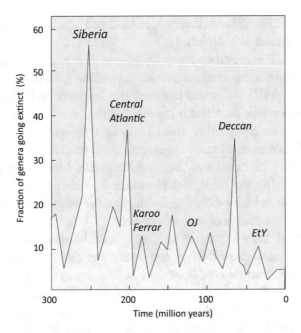

Fig. 3.3 Temporal relationship between the emplacement of Large Igneous Provinces and global faunal extinctions during the Phanerozoic. Extinction intensity is shown on the vertical axis as per cent (%) of extinct genera. OJ = Ontong Java Province (South-Western Pacific); EtY = Ethiopia-Yemen Province

kilometres. Such events release huge quantities of CO_2, H_2S, SO_2 and other gases into the atmosphere, producing toxic effects on both terrestrial and marine biota. There have been many LIP eruptions during the Earth's history, but those occurred during the Phanerozoic are associated with—and probably were the cause of—great mass extinctions[6] (Fig. 3.3).

The most voluminous of these LIPs formed about 200 million years ago, between the Triassic and Jurassic, before the fragmentation of the Pangaea supercontinent (Chap. 9). Lava and pyroclastic rocks, initially deposited on a surface more than 10 million km^2 wide, are now found over vast areas around the Central Atlantic Ocean, from northwestern Africa and southwestern Europe to North and South America.

Other important LIPs were emplaced in Siberia (250 Ma ago), the Karoo and Ferrar regions of South Africa and Antarctica (183 Ma ago), the southwestern Pacific Ocean floor near the Ontong Java atoll (110 Ma ago), the

---

[6] Courtillot and Olson [4], Sobolev et al. [5], Bond and Grasby [6].

Deccan area of India (67–60 Ma ago), Ireland and Scotland (60–55 Ma ago), and Ethiopia-Yemen (50–40 Ma ago).[7]

The Siberia LIP is coeval and causally related to the largest mass extinction ever documented on the Earth, which saw about 60% of the genera and 90% of the animal species disappear between the Permian and Triassic periods. The Deccan magmatic province is coeval with the extinction of non-bird-like dinosaurs and numerous other animal species between the Cretaceous and Palaeocene. This mass extinction is generally attributed to the fall of a gigantic meteorite in the Chicxulub area, Yucatan Peninsula, Mexico; however, recent studies indicate that the biological regression had already started before the meteorite impact, which probably gave the *coup de grace* to the animal species already in decline because of volcanism.

The origin of LIPs is one of the many controversial issues in modern geology. Some researchers call for the occasional rise of large hot blobs or **plumes** from the mantle-core boundary (D" layer) to the base of the lithosphere; here, hot rocks would undergo extensive melting, generating huge quantities of magma. According to other authors, such magmas come from thermally or chemically anomalous areas of the upper mantle.[8]

3.5 Global Distribution of Magmatism

Magmatism has been going on continuously throughout Earth's history. The oldest igneous rocks yet found are in Greenland and Quebec (Canada); they date back to Hadean time, more than 4 billion years ago, and have been profoundly modified by subsequent metamorphism. Additional evidence of very old magmatism is provided by the accessory mineral zircon contained in some rocks of Jack Hills (Western Australia). These minerals are magmatic in origin and have ages of 4.4 to 4.2 Ga, thus representing the oldest terrestrial material yet found (Chap. 9).

Proterozoic magmatic rocks are more numerous and better preserved. The Paleoproterozoic rocks from Bushveld, South Africa, which date around 2 Ga ago, are among the most famous. These make up an immense intrusive complex (i.e. an old magma chamber), cropping out over an area of several thousand square kilometres. Another famous intrusion, the 2.7 Ga old Stillwater Complex, is in Montana (USA).

---

[7] Siberia LIP erupted some 3–4 million km^3 of magma; Karoo-Ferrar one million km^3 magma; Ontong Java about 8 million km^3 emplaced in a submarine environment; Deccan nearly half a million km^3; the North Atlantic some 2 million km^3.

[8] Foulger [7].

The LIPs described above are the most significant occurrences of Phanero-zoic magmatic rocks. Magmatism associated with the so-called Hercynian (or Variscan) mountain building event (**orogeny**) from 380–280 Ma ago is also widespread across Europe (in Sardinia, Spain, Alps, Massif Central, Corn-wall, Bohemian Massif, etc.). Mesozoic-Cenozoic igneous rocks are common along the Alpine-Himalayan mountain chain, the Andes, the North America Cordillera, and many other places.

The number of volcanoes presently active on Earth is unknown. Historical eruptions are documented at about 1,500 centres; a presumably much greater number has occurred at submarine volcanoes, especially along the **mid-ocean ridges**, the nearly 65,000-km-long submerged volcanic chain running on the floor of the ocean basins.

The distribution of active volcanoes is shown in Fig. 3.4. It extensively overlaps with the distribution of seismicity, as discussed in Chap. 5. Most of the volcanoes are aligned along the edges of the Pacific Ocean (colloqui-ally known as the **Ring of Fire**), along the mid-ocean ridges, and in areas of continental rifting such as the East Africa Rift Valley and the Red Sea. Many isolated volcanoes occur in the middle of the oceans or continents (e.g. the Hawaii, Galapagos, Reunion islands in the Indian Ocean, Nyos in Cameroon, and many others).

The reason for the distribution of active volcanism will be discussed in Chap. 6. Here, it is sufficient to know that magmatism along ocean ridges and continental rifts is linked to the rise and melting of large masses of hot rocks from the convective mantle (Box 3.1). The volcanoes of the

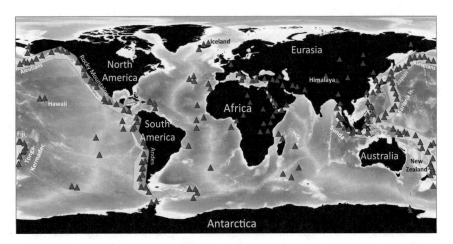

Fig. 3.4 Global distribution of recent and active volcanism. The yellow shaded band highlights the Circum-Pacific 'Ring of Fire'. The topographic base map is from NASA's website

Ring of Fire, Indonesia, Lesser Antilles, Aeolian Archipelago (Italy), and the Cyclades (Greece) originate from the hydration of upper mantle peridotite above subduction zones. In this geodynamic setting, the sinking lithosphere carries large amounts of water into the upper mantle. When released from the subducting lithosphere, this water triggers the partial melting of mantle peridotite, forming magma with a predominantly basaltic to andesitic composition. By contrast, the isolated intra-plate volcanoes are linked to thermal and/or chemical anomalies at limited places inside the upper mantle or the fusion of small hot plumes rising from the D" layer.

3.6   Summary

The history of the Earth can be viewed as the gradual evolution of a chaotic sphere of molten material into a layered planet that has been structured into a core, mantle, crust, hydrosphere, and atmosphere. Magmatism has played an essential and predominant role in this global modification; its action has consisted of the extraction of enormous masses of silicate liquids and gases from the interior of the planet and their transportation to the crust, hydrosphere, and atmosphere. Contemporaneously, heat loss to space has drastically lowered the Earth temperature.

Basalts are the most abundant magmas on Earth. They originate in the upper mantle by the partial melting of peridotite. Other abundant magmas include granite and andesite that make up a large fraction of the continental masses.

Volcanism is the external manifestation of magmatism. Gas emissions from volcanic centres impact the composition of the atmosphere, contributing to the preservation of the Earth's long-term optimal climate. The emission of volcanic CO_2 and the consequent greenhouse effect allowed our planet to get out of the numerous glaciations that have taken place in the recent and remote past. However, some large eruptions discharge huge volumes of lavas and gas in a geologically short time span. The effects of such phenomena have been catastrophic in the past and led to mass extinctions during the Phanerozoic.

The magmatic process, therefore, has contrasting effects on the Earth system and its biosphere. On the one hand, eruptions cause devastation, sometimes catastrophic and global, and on the other, magmas are the builders of the Earth's system—the only one known to host complex forms of life.

3.7 Box 3.1—Magmatism: How and Why

The temperature inside the Earth increases with depth, at a rate of about 30 °C per kilometre in the upper crust. Because of this phenomenon, one would expect that rocks deep inside the Earth to be in a liquid state. Instead, such a condition is rarely reached, because the pressure has the effect of compressing the rocks and raising their melting temperature. Moreover, the geothermal gradient does not keep constant, but diminishes significantly with depth (Chap. 1). As a result, the crust-mantle system—i.e. the silicate Earth—is almost entirely made up of solid rocks. The only place where the temperature nearly matches the melting point of rocks is in the asthenosphere, where some 1% liquid is dispersed among mineral interstices. However, these liquids are of insufficient volume to segregate and ascent to the surface.

If this description of the Earth is correct, where the magmas come from? The only liquid domain within the Earth is its outer core. However, a Fe–Ni composition cannot be the source of silicate magmas. Moreover, the metallic liquid is too dense to ascend to the surface.

Experimental studies of minerals and rocks at high pressures and temperatures show that the Earth's silicate interior may sometimes melt to produce magma. Melting occurs only in particular chemical-physical circumstances that may happen from time to time at localized sites in the upper mantle and the lower continental crust. The oceanic crust cannot melt because it is too thin and cold. Experimental studies also demonstrate that melting is never complete, but instead only affects a minor fraction, with a large portion of the rock remaining in the solid state. Summarising the results of numerous studies, rock of the upper mantle can partially melt only when at least one of the following conditions is fulfilled:

1—Decompression of extremely hot rocks. In particular places, large blobs of hot rocks may ascend from the intermediate or deep mantle, retaining their original heat content (adiabatic ascent); on reaching the base of the lithosphere, these rocks partially melt due to the drop in the pressure that lowers the melting temperatures of constituent minerals (**decompression melting**).

2—Arrival of fluids, especially H_2O, into the upper mantle. Such a process can occur in particular geological settings, such as above subduction zones, as will be discussed in Chap. 6. Water has the property of disrupting mineral structures at high pressure, thus drastically decreasing the melting temperatures (**hydration melting**). It has been experimentally demonstrated that an addition of 0.2% water to anhydrous peridotite in the asthenosphere lowers its melting temperature from about 1700 °C to 1250 °C.

Melting within the lower or intermediate continental crust primarily occurs when there are intrusions of basalt magmas from the underlying mantle. The arrival of hot liquids generates a rise in the temperature, favouring partial melting (Fig. 3.5).

Magmas exhibit a very wide range of compositions, most of which are rare and exotic (e.g. carbonatite and ultra-alkaline magmas). However, basalts, granites and andesites are the only ones to have a global distribution and a strong geological significance for the purposes of this book.

When the mantle or the crustal rocks undergo melting, some elements, called **magmatophile** or **incompatible**, preferentially go into the liquid phase, whereas others remain in the solid. This particular behaviour depends on the chemical characteristics of the elements, especially the atomic radius and valence. In general, the elements with a large atomic radius (e.g. potassium, rubidium, light rare earths, etc.) or high valence (e.g. thorium, uranium, niobium, etc.) as well as many gaseous substances, especially H_2O, cannot be hosted in the crystal lattice of the main rock-forming minerals, i.e. elements are incompatible with lattice structure of minerals. Therefore, they prefer to concentrate in the magma during the anatexis.

By contrast, other elements such as magnesium, chromium, nickel, and others prefer to remain in the solid mineral phases left behind by partial melting. Since the upper mantle is the refractory solid residue remaining after partial melting, the crust and the atmosphere-hydrosphere are more enriched in magmatophile elements and gases than the mantle, whereas the latter is enriched in Mg, Ni, Cr and other similar elements. The laws that govern

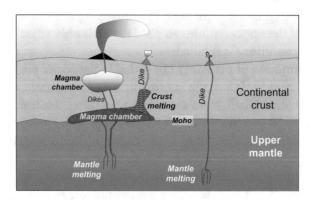

Fig. 3.5 Diagrammatic sketch of magma formation, segregation, and ascent. Partial melting of the upper mantle or the intermediate-lower continental crust generates basaltic and granitic magmas, respectively. Magmas rise towards the surface through fractures (dikes), sometimes erupting shortly after their formation or, in most cases, after resting for some time and partially crystallising inside crustal reservoirs (magma chambers)

the behaviour of elements in the magmatic systems were first discovered by Vladimir Vernadsky (1863–1945) and Victor Goldschmidt (1888–1947), the founders of Geochemistry and Crystal Chemistry.

A similar situation pertains to partial melting within the Earth's crust (Fig. 3.5). Therefore, granite magmas are more enriched in magmatophile elements with respect to the bulk crust.

3.7.1 Main Types of Magma

The number of magmas and the complexity of their chemical composition is an intricate maze, which is hard to cross, even for expert geologists. To sum up and simplify, **basalt** is the most abundant magma type on Earth and, more so, on the other terrestrial planets; it contains relatively low amounts of silicon, but is rich in magnesium, calcium, and iron. Volcanoes of the submarine mid-ocean ridges are the main eruption sites of basaltic magmas; however, numerous other basaltic eruptive centres are scattered around the world including the Hawaiian Islands, Etna, and the Galapagos. The cooling of basaltic magmas generates the effusive rock **basalt** and its intrusive equivalent **gabbro**; these rocks are dark in colour due to the presence of minerals rich in iron and magnesium, which are typically green or black coloured (Box 1.1). Basalts and gabbros make up almost the entire mass of the oceanic crust.

Granite is the second most abundant magma on Earth. It is generated by melting within the crust but can also derive from the chemical modification of basaltic magmas, as it will be briefly addressed later. After its formation, granite is emplaced at relatively shallow levels, constructing large sectors of the upper continental crust. Granite magma is rich in silicon and poor in magnesium, calcium, and iron. Rhyolites and granites are the effusive and intrusive rocks generated by the cooling of these magmas; both have fairly light-grey colours due to the dominance of white or colourless minerals, such as alkali-feldspars and quartz.

Andesite is the third most abundant variety of magma and is typically erupted at subduction volcanoes of the Circum-Pacific and Indonesian volcanoes as well as the Aeolian Islands (Italy), Cyclades Archipelago (Greece) and many other centres. Andesitic magmas have intermediate composition and physical properties between that of basalts and rhyolites. They can be formed by extensive melting of the lower crust or by compositional modifications of basaltic magmas in shallow reservoirs (**magma differentiation**). Along with granites, andesite magmatism has contributed substantially to the construction of the continents.

3.7.2   Ascent, Diversification and Solidification of Magmas

Magmas have a lower density than the rocks from which they form, and, therefore, tend to separate from them and rise towards the surface. Deep lithospheric fractures (**faults**) within the lithosphere are the channels along which magmas ascend to the surface. Faults are generated by extensional tectonic movements that break the lithosphere apart, allowing magmas to seep and move upward toward the surface. Although faults provide the pathways, the buoyancy force of magma is the reason for its upward movement. Ascent times can vary from hours to thousands of years. Some magmas, highly enriched in gas and therefore of very low density, can reach the surface a few hours after their formation. In most cases, however, the ascent is slow and interrupted one or more times by ponding within **magma reservoirs** or **chambers,** where magma can accumulate and reside for many years, centuries, or even millennia before the eruption.

Magma reservoirs generally develop inside the crust or along the **Mohorovičić discontinuity** (**Moho**) that separates the crust from the underlying mantle (Fig. 3.5). Their size is variable, reaching millions of cubic kilometres, in some cases. The place occupied by magma is the result of both extensional tectonic movements and the stress exerted by the magma itself, which creates its own space by pushing the solid rocks sideways and vertically. Once in the reservoir, magma loses heat, cools down, and partially crystallises before eventually ascending to the surface. In most cases, the residual magma solidifies completely within the magma chambers to generate **intrusive rocks**. Granites and other similar igneous rock are widespread around the world (e.g. Yosemite in the Sierra Nevada of California, the Cornubian granites in Cornwall, the granite mountains of the Scottish Highlands, Adamello, Mont Blanc, and other peaks of the Alps). These bodies were once deep magma chambers that solidified and subsequently were brought to the surface by tectonic uplift and erosion.

The solidification of magma within chambers is a slow process that goes on through the gradual crystallisation of different minerals and reduction in the amount of the liquid phase. The magma remaining after the crystallisation changes its chemical composition, becoming progressively enriched in silicon, alkali elements, other magmatophile elements, and water, and becoming depleted in concentrations of magnesium, calcium and iron. This process is referred to as **magma evolution** or **differentiation**. The result of prolonged crystallisation of a basalt magma is the generation of a number of derivative magmas, ranging from andesitic to rhyolitic-granitic composition. These various magma types can coexist in magma chambers and erupt at different times from individual volcanoes. A typical example is the Island of Vulcano, Southern Italy, where an early eruption of basaltic magmas (some 100,000 years ago) was followed by emplacement of andesite and finally rhyolite that all derived from the parental basalt by magma evolution.

3.8   Box 3.2—Heat Flow and Geothermal Energy

When descending into a mine or merely crossing through a deep road tunnel—such as the Gotthard Base Tunnel in Switzerland—it is apparent that the temperature underground is higher than on the surface, even during the hot summertime.

Accurate measurements along boreholes reveal that at depths exceeding some tens of meters, the temperature begins to rise at an average rate of about 30 °C per kilometre (the **geothermal gradient**). Geothermometric models indicate that such an increase remains fairly constant within the upper crust, but then decreases in the mantle and inner core, where the temperature gradient is around 1 °C/km or less. As a result, the temperature at the centre of the Earth is 'only' 5500–6000 °C, instead of 200,000 °C one would expect if the geothermal gradient remained constant throughout the entire Earth.

These observations indicate that the interior of the Earth is an enormous container of thermal energy. Simple calculations quantify the value to be around 10^{31} J, which is a billion times higher than the energy theoretically released by combustion of all of the world's oil reserves—thus it is an unlimited energy reservoir that stands under our feet, waiting for somebody clever who exploits it properly!

Heat is mostly produced by the decay of radioactive isotopes (e.g. potassium, uranium, thorium, etc.) occurring in the rocks, although there is some inherited primordial heat accumulated during the aggregation of the planet about 4.5 Ga ago. Other processes such as seismicity, rock movements inside the mantle and crust, crystallisation of the core and deformation of the crust under the effect of the lunar attraction (**terrestrial tides**) provide a much more modest contribution to the global thermal budget.

The Earth's internal heat continuously flows outwards, although the transfer mechanisms change from the core to the mantle and crust (Fig. 3.6). Heat removal from the inner to the outer core occurs by diffusion through the solid metal; the movements of the liquid metal in the outer core transport

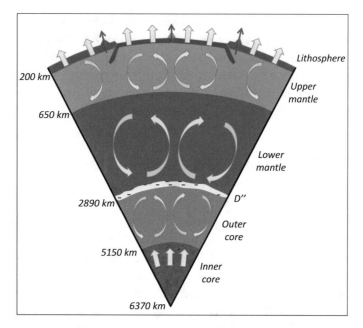

Fig. 3.6 Heat loss affects the entire mass of the planet and is accomplished by conduction in the inner core and the lithosphere (yellow arrows), and mass convection in the outer core and the mantle (circular arrows). Magmatism and hydrothermal activity are additional heat dissipation mechanisms in the lithosphere (thin red arrows)

heat to the D" layer and the lower mantle. The slow convective movements of the ductile rocks transfer heat from the lower mantle towards the transition zone and the upper mantle up to the base of the lithosphere. The final step of external heat dissipation occurs by conduction through the rocks of the lithosphere and by the ascent of hydrothermal fluids and magmas through rock fractures.

Diffusion through the rocks is the primary mechanism of heat transfer within the lithosphere. The average crustal heat flow is about 70 mW/m^2: an amount far too low to be of economic interest. Such a value means that capturing the energy released over one thousand square meters of the surface is necessary to power a 70-W light bulb.

However, in areas of recent or active volcanism, heat flow is much greater due to the presence of magma bodies under the surface at shallow depths. In these zones, geothermal energy is stored in hot rocks and fluids and can be conveniently exploited for the production of electricity and the heating of buildings or greenhouses. Interest in geothermal heat arises from low cost, unlimited abundance, and the moderate environmental pollution associated with exploitation. The first geothermal power plant was built at the beginning of the twentieth century in the Larderello area of Tuscany (Italy), where hot or still partially molten granite bodies sit at shallow depth within the crust. Numerous other plants have subsequently been established in zones of recent to active volcanism. The main world producers are the United States (around 6.7 billion kWh), followed by Indonesia (approximately 1.9 billion kWh), Philippines (about 1.9 billion kWh), Turkey (1.3 billion kWh), New Zealand (1.0 billion kWh), Mexico (951 MW), Italy (944 MW) and Iceland (755 MW).

3.9 Box 3.3—Magmatism and Ore Deposits

Construction of many everyday objects with which we are familiar, such as light bulbs, cars, trains, aeroplanes, mobile phones, and computers, utilize metal elements such as lithium, aluminium, titanium, iron, cobalt, copper, zinc, niobium, tantalum, tungsten, and the rare earths. Mineral resources are the indispensable building blocks of modern technology.

All of these elements are found in the Earth's crust. Their abundance, however, is on an average too low for it to be of any practical use. However, there are areas where chemical elements are concentrated in large quantities to form mineral deposits that can be profitably extracted. Most of these concentrations are genetically related to magmatic processes.

The accumulation of useful minerals can take place during magma crystallisation (**orthomagmatic** and **pegmatitic** deposits) or later on, by the action of the fluids that scavenge the elements from both magmas and adjacent heated rocks and concentrate them along cracks and voids (**hydrothermal** and **epithermal** deposits). The formation temperatures of the orthomagmatic and pegmatitic deposits range between about 1200–700 °C. Hydrothermal and epithermal deposits are accumulated at lower temperatures.

Orthomagmatic ore deposits can be generated by the crystallisation of minerals in magmas or by the unmixing of liquids of contrasting composition, such as silicates and sulphides. Important deposits of this type are chromium, platinum, nickel, and metal sulphides at Bushveld (South Africa) and Stillwater (Montana, USA).

Pegmatitic deposits originate in the latest stage of crystallisation of intrusive magmatic bodies (about 600–700 °C) when magmas are extensively crystallised. The residual silicate liquids are enormously enriched in water, silicon, alkali elements, and rare magmatophile elements such as lithium, beryllium, niobium, tantalum, the rare earths, uranium, and thorium. The solidification products of these melts are called **pegmatites**. These are some of the most spectacular rock types, being made up of large crystals, often with well-developed distinctive forms. Their economic interest is linked to both the high concentrations of rare elements and the presence of well-shaped crystals of quartz, tourmaline, aquamarine, topaz and many other minerals that are highly valued as gemstones and museum specimens.

Hydrothermal and epithermal ore deposits are genetically related to fluids separated from solidifying magmatic intrusions (H_2O, CO_2, HF, etc.). Such fluids extract elements from the solid but still hot magmatic body or the adjacent country rock, and concentrate them into rocks away from the intrusion. Hydrothermal minerals accumulate at a temperature of about 700–300 °C, generally around deep magmatic bodies. Epithermal deposits are generated at lower temperatures, between 300 and 100 °C, in sub-volcanic environments close to the surface.

Finally, many sedimentary ore deposits are secondary concentrations of magmatic minerals. These have been removed from their original place by erosion, and subsequently deposited in particular places by running surface waters to form accumulations of gold, platinum, diamonds, and various other minerals, which are referred to as **placers**.

References

1. Anderson DL (2006) New theory of the Earth. Cambridge University Press, 384 p
2. Briffa KR, Jones PD, Schweingruber FH, Osborn TJ (1998) Influence of volcanic eruptions on Northern Hemisphere summer temperature over the past 600 years. Nature 393:450-455
3. Saunders AD (2005) Large Igneous provinces: origin and environmental consequences. Elements 1:259–263
4. Courtillot V, Olson P (2007) Mantle plumes link magnetic superchrons to Phanerozoic mass depletion events. Earth Planet Sci Lett 260:495–504
5. Sobolev VS, Sobolev A, Kuzmin VD, Krivolutskaya NA, Petrunin AG, Arndt NT, Radko VA, Vasiliev YR (2011) Linking mantle plumes, large igneous provinces and environmental catastrophes. Nature 477:312–316
6. Bond DPG, Grasby SE (2017) On the causes of mass extinctions. Palaeogeogr Palaeocl 478:3–29
7. Foulger GR (2010) Plates vs plumes: a geological controversy. Wiley-Blackwell, 340 p

4

Geomagnetism—The Space Shield of the Planet Earth

Infiniti autem sunt meridiani magnetici, eodem etiam modo dirigentes se, per certos et oppositos in equatore terminos, et polos ipsos. In illis etiam latitudo magnetica mensuratur: et inde declinationes intelliguntur; et in illis directio certa in polos tendit, nisi malo aliquo variauerit, et de iusta via disturbetur magneticum. (The magnetic meridians, then, are numberless, arranged in the same direction, through fixed and opposite points on the equator and through the poles. On them the magnetic latitude is also measured; and by means of them the declinations are understood; and along them a fixed direction points towards the poles, unless there is some anomaly that deviates the magnetic body from the right way.)
William Gilbert, De Magnete (1600)

4.1 Introduction

Magnets are natural or man-made objects that have the power to attract iron and a few other metals. Such a property is related to a **magnetic field**, an invisible set of forces that envelope the magnet.

A well-known classroom experiment to reveal the existence of a magnetic field is sprinkling iron filings on a sheet of paper or a glass plate placed above a bar magnet; the iron particles concentrate at the extremes of the bar but also distribute along particular directions, which indicate the lines of force or vectors of the magnetic field (Fig. 4.1).

The extremities of the bar are called **magnetic poles**. Opposite poles are referred to as south (or positive) and north (or negative). Unlike poles of

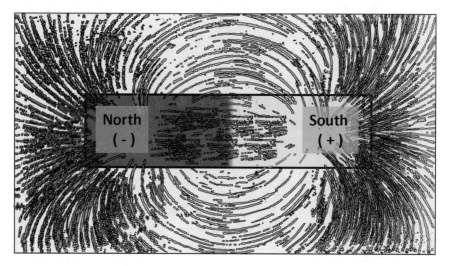

Fig. 4.1 Arrangement of iron filings around a magnetic bar. The array of iron particles follows the vectors of the magnetic field of the bar

magnets attract each other whereas like poles repel. Then, if two magnets are put side by side, the positive pole of one magnet will direct toward the other's negative pole.

Poles are always found in pairs, and the magnetic field is referred to as **dipolar**. Single magnetic monopoles have been proposed theoretically, but have never been observed. Therefore, if a magnet is broken into pieces, each of the fragments will become a dipole.

Magnetism is closely related to electricity. A moving electric charge generates a magnetic field. Conversely, a magnetic field is capable of producing an electric current (Box 4.1). Electromagnetism is one of the primary forces in nature, along with gravity, weak nuclear forces and strong nuclear forces.

Substances capable of creating a strong magnetic field exist in nature and can be constructed artificially. **Lodestone** is a natural permanent magnet; it is made up of **magnetite** (Fe_3O_4), a mineral diffused in small quantities in almost all rocks and sometimes concentrated in large deposits. Only some varieties have the power to attract iron and other metals, such as nickel, cobalt and certain rare earth alloys; however, all magnetite crystals are attracted by magnets.

Artificial magnets can be created by putting iron in contact with the lodestone; however, magnetisation is temporary. Permanent magnets can be constructed by applying a strong magnetic field to some special iron alloys during manufacture. This treatment induces an alignment of constituent atoms, which are small dipoles and, if arranged in the same direction, impart a permanent magnetism to the metal.

The intensity of the magnetic field of lodestone or other material decreases from room temperature to the so-called **Curie point**. This is the critical temperature at which materials lose their permanent magnetic properties; its value changes from one substance to the other.[1]

4.2 The Earth's Magnetic Field

If a splinter of lodestone, or an artificial permanent magnetic needle, is left free to move—for example, by hanging it to a cotton thread—it will naturally rotate until it aligns with the Earth's meridian, with one end pointing to the north and the other to the south. This property is the principle on which the **magnetic compass**—one of the most important instruments for navigation –is based. The end of the compass needle pointing northward on the Earth is called the north pole of the magnet; the opposite one is referred to as the south pole of the magnet.

The spinning and the north–south alignment of a compass needle provide the most simple and effective evidence that Earth is surrounded by a magnetic field. In other words, the Earth behaves as a large magnetic bar whose south pole attracts the north pole of the needle and repels the south one. The ideas on the causes of the Earth's geomagnetic field have evolved with time and will be briefly discussed in Box 4.1.

The magnetic field near the Earth's surface has a complex structure, but can be illustrated in a very simplified manner as a set of vectors (**magnetic field lines**) that come out of the south terrestrial pole region and enter the north one after following elliptical paths around the Earth (Fig. 4.2a). This configuration is the same as that of a magnetic field generated by a hypothetical magnetic bar sitting inside the Earth.[2]

The axis of the magnetic dipole is inclined with respect to the Earth's rotation axis by an angle that varies continuously over time and is currently slightly less than 10 degrees. Therefore, the two emergence points of the magnetic field axis on the Earth's surface—called **geomagnetic poles**—are close, but do not coincide with the geographical poles.

---

[1] The Curie point (from Pierre Curie, who studied magnetism) for iron is 769 °C, and that of magnetite is 585 °C.

[2] Since the north pole of the compass needle is, by definition, the end pointing north, the south magnetic pole of the bar – that attracts the northern pole of the compass - has to be thought as situated in the northern hemisphere of the Earth, as indicated in Fig. 4.2a.

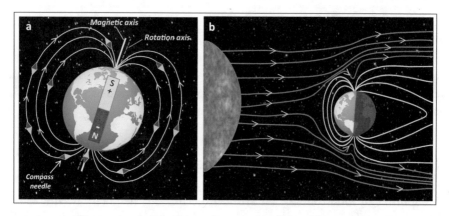

Fig. 4.2 **a** Simplified 2-dimensional image of the magnetic field around the Earth; **b** Deformed magnetosphere (yellow lines) under the pressure of the solar wind (red lines with yellow arrows). The penetration of the solar wind particles at the two poles ionizes the atmospheric gases that produce the brightness of the polar auroras

The strength or **intensity** of the geomagnetic field on the Earth's surface is variable and increases from some 30,000 nanotesla (nT) at the equator to about 60,000 nT at the poles.[3]

The volume around the Earth where the magnetic field is active and the charged cosmic particles are affected by geomagnetism is called **magnetosphere**. Its overall shape is strongly elongated, with a thickness of about 65,000 km on the side facing the Sun (dayside) and a long tail that extends for millions of kilometres in the opposite direction (Fig. 4.2b). The strong asymmetry of the magnetosphere is the consequence of the pressure exerted by the solar wind, a diluted plasma of protons, electrons and alpha particles that are continuously emitted by the Sun and diffused in the space, blowing against the magnetosphere of the Earth and other planets. The solar wind particles are charged and contain a magnetic field; they impact upon the magnetosphere, modify its shape and are finally deflected around it to disperse in the open space. Therefore, the magnetosphere inhibits the solar wind particles to penetrate the atmosphere and reach the Earth's surface.

Solar wind particles have destructive effects on cells of living organisms and the molecules of atmospheric gases that are split into their atomic components and dispersed in space. Without the presence of a strong magnetic field, the implacable action of solar particles would kill organisms and strip the planet of its atmosphere. The latter process operates effectively on Mars

---

[3] The tesla (T) is the unit of measure in the *Système International (SI)* for magnetic induction. The nanotesla (nT) is equal to one billionth of a tesla. Another unit of measurement is gauss G (1 G = 10^{-4} T).

where the magnetic field is only a few nanotesla in strength. The magnetic field, therefore, is a space shield that effectively protects the Earth and its inhabitants. Without it, there would be little or no atmosphere, hydrosphere, and life. Interestingly, geological evidence suggests that the geomagnetic field was active by at least 3.7 billion years ago, around the time early life appeared on Earth.[4]

The trend and intensity of the geomagnetic field can be identified by a **magnetometer**. This instrument is presently used for a wide variety of applications, from sensing the geomagnetic field to detecting metallic objects hidden below the ground or under the water, or anything else that causes a local magnetic perturbation.

There are several types of magnetometers that employ different detection methods. The simplest variety is a compass, consisting of a magnetic needle suspended by a thread or resting on a pin, free to swing in space. At any place on the planet, the needle will align itself with the geomagnetic vector at that particular point. A small displacement will make the needle oscillate around the equilibrium position. The frequency of oscillations depends on the intensity of the magnetic field.

Owing to the geomagnetic field array, the magnetised needle of the compass takes a different position at various places around the Earth (Fig. 4.2a). In particular, the angle with the horizontal plane (**magnetic inclination**) varies from zero degrees at the magnetic equator (horizontal needle) to 90° at the magnetic poles (vertical needle); the angle with the terrestrial meridians (**magnetic declination**) varies between zero and 180°, as a consequence of the different positions of the magnetic and geographical poles (Fig. 4.3a).

Based on the above, it might seem obvious that the latitude and longitude of any location on the Earth's surface can be identified by the inclination and declination of a compass. The latitude can also be somewhat deduced from the intensity of the geomagnetic field. However, the procedure, although conceptually correct, is much less straightforward than indicated above. Difficulties arise from various factors, including continuous changes in the shape and strength of the Earth's magnetic field over time, local and regional anomalies, and the occurrence of non-dipolar components. These factors make the near-surface geomagnetic field complex and resembling much more a tangled ball of yarn than the orderly arrangement of magnetic forces shown in Fig. 4.2a.

---

[4] Witze [1].

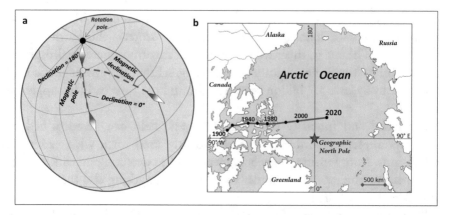

Fig. 4.3 **a** The magnetic declination is the angle between the directions of magnetic north (indicated by the compass needle) and geographic north (indicated by the meridian). It can vary from 0° to 180°; **b** Migration of the north magnetic pole during the last century. The speed of 0–15 km/year between 1990 and 2005 has accelerated to the present rate of 50–60 km/year

The Earth's magnetic field is subjected to continuous and sometimes radical modifications.[5] One of the most interesting secular variations is the migration of magnetic poles that continually change their position, although remaining relatively close to the geographical poles. Measurements made in the last century indicate that the north magnetic pole has moved some 2000 km away from the Taloyoak area (Nunavut Territory, Canada), towards the Arctic Ocean and Siberia (Fig. 4.3b). The south magnetic pole has also migrated from Victoria Land to offshore Terre Adélie.

The intensity of the magnetic field is also changing continuously. In the last two centuries, since the invention of the magnetometer, the magnetic field intensity at the Earth's surface has decreased by about ten per cent.

4.3 Palaeomagnetism: The Magnetic Memory of Rocks

Modern techniques allow recognising the characteristics of the Earth's magnetic field, even for the past geological eras. This is possible because rocks contain a weak but permanent 'remaining' or 'fossil' magnetism acquired during their formation from the surrounding Earth's magnetic field. The inclination, declination, and intensity of the fossil magnetic field in the rocks

---

[5] http://www.geomag.bgs.ac.uk/education/earthmag.html
    Hulot et al. [2].

record the features of the geomagnetic field existing at the time and place the rocks were formed. Since inclination, declination, and intensity depend on latitude and longitude, their determination allows one to deduce the geographical position of the rocks at the time of their formation. The study of fossil magnetism or **palaeomagnetism** has provided geologists with some of the most valuable information on our planet's past, which has had a decisive impact on our understanding of the Earth's evolution.

Fossil magnetisation is measurable only with sophisticated instruments and procedures. However, it can sometimes be crudely detected by placing a compass near or onto an ancient rock, especially a basalt: if the declination of the fossil magnetism in the rock differs significantly from the current one, the needle rotates quickly to align with the magnetic field preserved in the rock.

Fossil magnetism is linked to ferrimagnetic minerals present in a rock, especially magnetite. However, the mechanism by which remnant magnetism is acquired differs from sedimentary to magmatic processes.

Remanent magnetism in sedimentary rocks is a consequence of the deposition of magnetite granules during sedimentation in aqueous environments. Magnetite crystals settle down taking a preferred orientation, roughly parallel to the surrounding magnetic field lines, basically behaving like compass needles. Once embedded in the rock, the granules remain blocked, preserving the record of the geomagnetic field existing at the time and place of sedimentation. By contrast, fossil magnetism in igneous rocks is linked to the tiny granules of magnetite that crystallised in the magma during cooling. When magma turns into rock and the temperature drops below the Curie point of magnetite (about 585 °C), the mineral granules acquire a magnetisation that lines up with the ambient magnetic field.

Over the past 70 years, paleomagnetic studies in both continental and marine environments allowed geologists to demonstrate that ancient rocks were formed far away from their present geographical position. The data provided an irrefutable demonstration that crustal rocks making up the continents and ocean floor are not immobile, but have instead moved horizontally over long distances during the geological past.

4.3.1 Magnetic Reversal

One of the most astonishing results of palaeomagnetic studies is the discovery of **magnetic reversal**, i.e. the interchange of the Earth's magnetic field polarity, with the geomagnetic north switching to the south, and *vice*

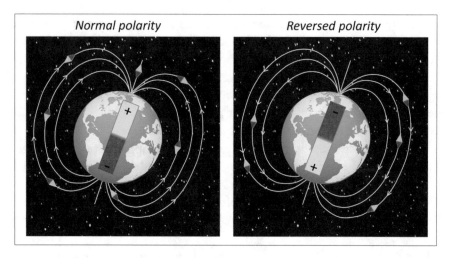

Fig. 4.4 The Earth's 'normal' and 'reverse' magnetic fields. During 'normal' stages, the geomagnetic field can be viewed as related to a magnetic bar occurring inside the Earth: the south pole of the bar is sited in the northern hemisphere, attracts the north pole of the compass needle and repels the south one. An opposite setting can be envisaged for the reverse stages

versa. The phenomenon was first discovered in 1906 by French geophysicist Bernard Brunhes (1867–1910) while studying lava flows from the France Massif Central.

The current polarity is defined as '*normal*' and the opposite one '*reversed*' (Fig. 4.4). A single period, with a prevailing normal or reverse polarity, is called **chron**. The geological record indicates that single chrons can last almost any length of time, from tens of thousands to millions of years, and there is no apparent pattern in their duration. Periods shorter than 200,000 years are referred to as **subchrons**.

Magnetic reversals have occurred hundreds of times throughout the history of Earth. The last reverse magnetic polarity chron (Matuyama) started 2.6 million years ago and ended 780,000 years before the present, passing to the current normal chron (Brunhes). Although reverse polarity prevailed during the Matuyama chron, there were many subchrons with normal polarity, some of which lasted only a few tens of thousand years.

The switch from one chron or subchron to the next takes a few hundred to thousands of years to attain completion. However, there is evidence for much more rapid reversal times, sometimes comparable to average human life.[6] The reversal starts with a strong attenuation by 80–90% of the magnetic field

---

[6] Sagnotti et al. [3].

intensity and the dipole deformation into a quadrupole before the complete flip. Some geophysicists speculate that the decline of the intensity and the acceleration of polar migration suffered by the Earth's magnetic field during the last century (Fig. 4.3b) may be the first signs of an upcoming reversal.

Many numerical models qualitatively reproduce the structural and dynamic characteristics of the geomagnetic field, including reversals. Yet, the precise reversal mechanism remains controversial. Some scientists believe that periods of higher turbulence inside the liquid outer core cause the magnetic field to destabilise and change the polarity. Others suggest that convection of liquid outer core is naturally prone to instability, but a switch of polarity is hindered by the stabilising effect of the solid inner core.

The effect of magnetic reversals on the biosphere is not clear. Some palaeontologists emphasise the disappearance of certain living species contemporaneously with a magnetic switch, suggesting reversal as the cause of extinctions. Others believe there is no compelling evidence for a causal relationship between the two phenomena (Box 4.2).

4.4 Palaeomagnetism, Continental Drift, and Ocean Floor Spreading

The study of palaeomagnetism has contributed decisively to the birth of the most revolutionary theories in the field of Earth Sciences: the continental drift and the spreading of the ocean basins, from which the scientific paradigm of plate tectonics germinated (Chap. 6).

The idea of continental drift, grasped by some scientists as early as the sixteenth century, was originally developed and scientifically documented outside the earth sciences community by the German meteorologist Alfred Wegener (1880–1930), at the beginning of the twentieth century, as explained in Chap. 6. The theory encountered fierce opposition, discredit and rebuttal by many geologists, including Sir Harold Jeffreys (1891–1989), an outstanding geophysicist who was the first to demonstrate that the Earth's outer core is liquid. However, palaeomagnetic studies carried out by American and British geophysicists in the 1950s found that the orientation of the palaeomagnetic field measured in rocks in Britain and North America was strikingly different from the present one.[7] Similar results were successively found for many rocks around the world.

---

[7] Creer [4], Runcorn [5].

These findings could be interpreted as evidence that the magnetic poles had been wandering over the Earth surface through geological times. However, curves of magnetic polar migration for different continents did not match, as expected if poles were moving and continents were fixed. Paleo-magnetic evidence could only be explained by assuming that the continental masses had shifted away from their original positions, each in a different direction. Such a conclusion was a decisive blow to old ideas of a fixed and immutable position of the Earth's crust (**fixism**).

An additional thrust to crustal mobility theories (**mobilism**) came from the study of the ocean floor. The need to map the seafloor topography in detail, essentially for military purposes during the Cold War, prompted systematic oceanographic surveys that greatly expanded our knowledge about regions of the planet that were, until then, substantially unknown. The most extraordinary result of oceanographic surveys was the discovery of the **mid-ocean ridge**, the enormous underwater chain of volcanoes, a few thousand kilometres wide and a few thousand metres high, which stretches continu-ously on the ocean floor over a distance of about 65,000 kms. It does not always run in the middle of oceans, but this happens in the Atlantic Ocean where the structure was first identified and named.

Geophysical studies and rock samplings showed that the oceanic ridges are mostly made up of basalts, covered by a thin and irregular veneer of sediments. The thickness of the sedimentary layer increases away from the ridge towards the surrounding abyssal plains but never exceeds a few hundred metres. This means that sediments did not have enough time to pile up thick sequences, basically speaking for a relatively young age for oceans and seafloor rocks.

Magnetic surveying and radiometric age dating provided additional amazing results. It was observed that the central sectors along the axis of the ridges consist of very young rocks, which gradually become older towards the adjacent abyssal plains, on both sides of the ridge (Fig. 4.5a). It was also observed that the magnetic polarity is 'normal' for young rocks along the ridge axis, but changes several times and symmetrically between 'normal' and 'reverse' on both the flanks of the ridge, in parallel with the increasing ages of rocks. Overall, seafloor consists of parallel strips of rocks with alter-nating polarities symmetrically distributed on the two sides of the ridge, as two mirror images.[8] This chronological and geomagnetic symmetry was best demonstrated for the Reykjanes Ridge, offshore southwest Iceland (Fig. 4.5b).

---

[8] The magnetic survey was carried out by towing a magnetometer a few hundred meters astern the ship to minimize the effects of the vessel magnetic field. The sensor was kept at the height of a few meters above the seabed and was drifted slowly across the rock strips, permitting maximum

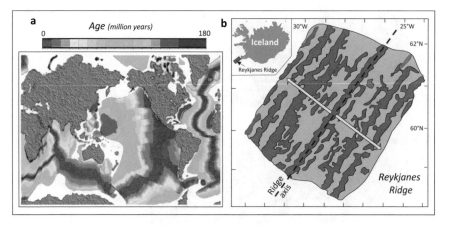

Fig. 4.5 **a** Distribution of seafloor rock ages at the sides of mid-ocean ridges. The deep-red colour indicates the younger ages along the axis of the ridges; the light-red, yellow, green and blue colours denote progressively older rocks. The oldest ages for the present oceans (about 180 million years) are found near continental margins (modified after a NOAA map); **b** Symmetrical distribution of magnetic polarities in the rocks of the Reykjanes Ridge, south of Iceland. Red stripes are rocks with normal polarity; green stripes denote rocks with reverse polarity. Yellow arrows indicate the direction of spreading and the increase of rock ages away from the ridge axis

The geochronological, petrological, and magnetic data led to the conclusion that the axial zones of the oceanic ridges are areas of crustal extension and eruption of basaltic magmas. The extensional tectonic activity displaces the rocks away from the rift axis, generating gaps along which mantle-derived basaltic magmas intrude and erupt onto the seafloor. Continuous intrusion and solidification of magmas into cracks and eruption on the seafloor create new oceanic crust. The lavas presently outcropping in the centre of the ridges are the youngest and, therefore, have normal magnetic polarity; the rocks located on the sides of the ridge become progressively older and show an alternating normal or reverse polarity, depending on the characteristics of the geomagnetic field existing at the time of their eruption. Therefore, oceanic ridges are the sites of continuous crustal extension, eruption of basaltic magmas, formation of new crust and lateral displacement of crustal blocks: in short, a process of seafloor spreading that, over time, opened and enlarged the ocean basins (Fig. 4.6).

Like many innovative theories, the hypothesis of oceanic opening proposed by Lawrence Morley, Fred Vine, Drummond Matthews, Robert Dietz, Harry Hess, and others found hostility by the scientific community. It is reported

definition of the shape and amplitude of magnetic variations. Airborne measurements can also carry out magnetic surveys. Heirtzler et al. [6].

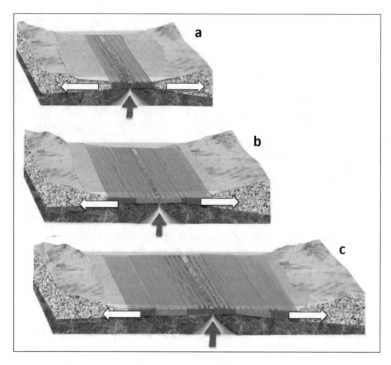

Fig. 4.6 Opening of an ocean by spreading, and variation of the magnetic polarity of volcanic rocks. **a** The extensional fracturing of a continent gives rise to a rift valley and the eruption of volcanic rocks showing the same magnetic polarity as the surrounding geomagnetic field (for example, normal polarity in red colour); **b** Continuing spreading promotes the emplacement of new lavas, whose magnetic polarity may be reverse (deep green rocks); **c** Long-term expansion causes the formation of an ocean floor comprised of parallel rocks stripes with alternating normal and reverse polarity, distributed symmetrically on the two sides of the ridge—a kind of recorder of Earth's magnetic field during the various stages of ocean basin expansion

that a referee of a prestigious scientific journal judged the ideas on seafloor spreading as interesting speculations for cocktail parties but not the sort of things that one expects to read in the qualified scientific literature.[9]

However, scepticism and opposition must not be blamed. In fact, if the oceanic expansion theory explained the magnetic features and the ages of rocks on the seafloor, it also raised new and more puzzling problems: Did oceanic expansion imply that the planet volume is becoming larger with time? Is there any evidence for planet growth? Or is there some unknown process that destroys the crust created along the ridges? And where does such a destruction take place? The answers to these questions, which finally put

---

[9] Stern [7].

scepticism to rest, were provided shortly by the plate tectonics theory, which will be discussed in Chap. 6.

4.5 Summary

The Earth is enveloped in a strong magnetic field whose shape can be represented as a set of vector lines that come out of the magnetic poles and draw elliptical paths in space. Viewed from afar, the region of space around the Earth, where the magnetic field is active (**magnetosphere**), shows an asymmetrical shape, with a sunward thickness of about sixty thousand kilometres and a long tail that extends for millions of kilometres in the opposite direction. Such an enormously elongated geometry is produced by the movement toward Earth of the solar wind, a diluted plasma of charged particles (mostly protons) emitted by the Sun that is continuously impacting our planet and other solar system bodies.

The solar wind can strip the atmosphere off planets. However, the magnetic field acts as a shield that deflects the charged particles into space, inhibiting their penetration into the atmosphere. The geomagnetic field, therefore, is a powerful protective shield against solar and cosmic radiations and is a decisive factor in the conservation of the atmosphere and hydrosphere, and the development of life on Earth. Ablation of the atmosphere has been actively working over geological time on Mars, where the present geomagnetic field is very feeble.

The Earth's magnetic field is subjected to several modifications over time scales of days to millions of years. The migration of the magnetic poles with respect to the geographical ones is one of the most remarkable secular variations. Much more radical changes occur during the reversals of the magnetic field's polarity, when the magnetic north switches to the south, and *vice versa.* The Earth's magnetic field has flipped hundreds of times over its history.

The modifications of the Earth's magnetic field during geological times have been revealed by palaeomagnetic studies, i.e. by investigations of fossil magnetism acquired by rocks at the time of their formation and preserved until the present time. Palaeomagnetic data have provided irrefutable evidence for demonstrating that the Earth's crust is mobile and the oceans are expanding, establishing the foundations for plate tectonics theory, the present-day over-arching scientific paradigm for the Earth Sciences.

4.6   Box 4.1—Geomagnetism: A Historical Perspective

The properties of lodestone (meaning "leading stone", the stone that indicates direction) have been known since ancient times and attracted the attention of many philosophers and naturalists. Ancient Greeks located the source of the stone in western Asia Minor (presently Turkey), near the city of Magnesia, where the word magnetism comes from.

Thales, a Sage of the ancient Greek colony of Miletus, believed loadstone was a kind of animated material possessing a sort of soul that attracted or repelled other objects—a conception that lasted for centuries. The Latin naturalist and poet Titus Lucretius Caro (98–55 BCE) in his poem *De Rerum Natura* (On the Nature of Things) talks about science and religion, discrediting superstition and myths in interpreting nature. Lucretius excludes the lodestone from having a soul and instead suggests that atoms emanate from the stone, push the surrounding air away, and create a vacuum that allows the iron to be suctioned toward the stone. Apart from the interpretation of loadstones, many ideas of Lucretius were amazingly ahead of his times and can be still acceptable today.

The Chinese in the third century BCE were the first to use lodestone to make compasses. However, their original use was not for travelling, but rather for aligning the building, believing that houses placed in the right direction would bring a healthy and wealthy life to their owners—a practice known as *feng shui*. However, the use of compasses for travelling and navigation did start in China, but later at around the beginning of the second millennium of the common era. A couple of centuries later, the compass arrived in Europe through caravan trade routes from the east and soon became an essential navigation tool.

The first systematic description of magnets is found in *Epistula de magnete* (Letter on Magnet) a letter on lodestone written to a friend by the French scholar Petrus Peregrinus (Pierre Pelerin de Maricourt) in 1269, while he was in Italy besieging the town of Lucera with the troops of Charles, the duke of Anjou. The author illustrates the properties of the compass, speaks about the opposite poles of lodestone, and states that these are preserved even if the stone is crushed, since all splinters maintain the same dipolar characteristics as the original material.

The Venetian Jesuit Leonardo Garzoni (1543–1592) performed experimental studies on magnetism, described in his work *Due Trattati sopra la natura e le qualità della calamita* (Two Treatises on the Nature and Quality of Lodestone). He carried out 90 experiments that demonstrated, among other things, the impossibility of shielding magnetic attraction and falsified some old ideas, including those that attributed the phenomenon to effluvium or the suction of air by magnetite. Interestingly, Garzoni's work was printed only a few years ago, but it had a large circulation as a manuscript in his day.

The idea that the Earth could be considered a colossal magnet surrounded by what we now call a magnetic field, emerged in the book *De Magnete* (On the Magnet), by William Gilbert (1544–1603), the first real scientific work on magnetism. Gilbert distinguished the attraction of magnets from that of amber. He performed an interesting analogue experiment by building a magnetite sphere, a scale model of the Earth or *terrella* (small Earth), which, among other things, helped him reconstruct the array of the Earth's magnetic field.

According to Gilbert, the Earth's interior contained a large magnetised metal bar, an idea abandoned when it became clear that this body would be subjected to high temperatures, exceeding the so-called Curie point for any known substance. Moreover, the magnetic bar hypothesis implied that the Earth had a stable magnetic field. This was falsified by the work of Carl Friedrich Gauss (1777–1855), who demonstrated that the geomagnetic field is not a perfect dipole and changes continuously over time, something that could only be explained by assuming that electric currents generate magnetism.

Advancement in understanding magnetism was gained thanks to well-known experiments by the Danish Hans Christian Oersted (1777–1851), and the English Michael Faraday (1791–1867). It was shown that a straight copper wire crossed by an electric current generates a magnetic field that turns the compass needle in an orthogonal direction to that of the wire; *vice versa*, the rotation of a coil of conducting wire around a magnet, like in a dynamo, generates an electric current.

Eventually, it became evident that electricity and magnetism were two closely related phenomena. The most outstanding contribution to this outcome was given by the Scotsman James Clerk Maxwell (1831–1879), the great theoretical physicist of the nineteenth century. He made an ingenious synthesis of the work of Oersted, Coulomb, Gauss, and Faraday to develop the unifying theory of electromagnetism. In his famous four equations, he demonstrated that electric and magnetic forces are not two separate phenomena, but different manifestations of a single electromagnetic force.

Coming back to the magnetism of the Earth, the present view is it results from both internal and external factors. The endogenous, or intrinsic, component is largely dominant. Paleomagnetic record indicates that the geomagnetic field has existed for at least 3.7 billion years. Such a field would decay in only a few thousand years if not continually being generated by some kind of endogenous process. The mechanism responsible for fuelling the geomagnetic field is the so-called **dynamo effect**, a self-exciting system powered by the motion of liquid iron inside the outer core (Fig. 4.7).

The dynamo theory was proposed for the Sun in 1919 by the Irishman Joseph Larmor (1857–1942) and underwent subsequent refinements by the German scholars, Walter Elsasser (1904–1991) and Friedrich Busse of the University of Bayreuth, Anglo-American Edward Bullard (1907–1980), and many others. The self-sustained dynamo mechanism works inside the Earth but requires an initial magnetic field in the primordial liquid core to start. Conductive metal moving across the magnetic field generates electric currents that, in turn, give rise to a secondary magnetic field that couples with the initial field, making it stronger and roughly orienting it along the rotational

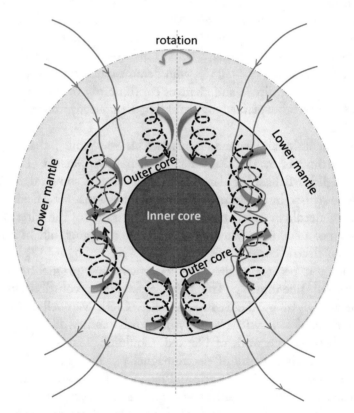

Fig. 4.7 Simplified picture of the Earth's liquid core dynamics that generates its magnetic field. Compositional and thermal gradients in the liquid metal promote convection (large red arrows); the Earth's rotation causes swirl-type movements (dashed lines) parallel to the rotation axis, which act as a dynamo to generate the geomagnetic field

axis of the planet. Once the dynamo is started, the process goes on indefinitely, being sustained by the liquid mass movements. At present, the dynamo mechanism is working in the outer core of the Earth. The heat released from the continuously crystallising inner core (**latent heat of crystallisation**), and the thermal dispersion ensured by the overlying convecting mantle, along with impurities left behind by the solidifying metal, generate thermal and compositions gradients in the outer core, which induce convection. Additional movements are caused by Coriolis forces related to the rotation of the Earth (Fig. 4.7).

The Earth is not the only planet whose magnetic field is related to a dynamo. The same mechanism works in the Sun and some planets of the solar system, except Mercury, Venus, and Mars. The lack of a strong intrinsic magnetic field in these planets is probably related to the relatively small size of Mercury and Mars that can hold only a very small convective core. Venus has similar bulk characteristics as the Earth, but rotation around its axis is about 250 times slower than Earth, probably hindering the dynamo mechanism. The Sun has a very energetic dynamo mechanism that generates a reversal of its magnetic field approximately every 11 years.

4.7   Box 4.2—Geomagnetism and the Biosphere

There is a wealth of evidence that the Earth's magnetosphere affects living organism, from bacteria to plants and complex animals. It has been experimentally demonstrated that some bacteria in ponds and lakes swim in the direction of the Earth's magnetic field. When these bacteria are brought into an artificial magnetic field, they adjust their direction of movement according to the new vectors of magnetic forces. Bees use sunlight as a compass to reach their feeding places and return to apiaries, but they also feel the magnetic field. A magnetic compass is especially useful during overcast days when the other mechanisms of navigation are not available.

Many fish, such as trout, salmons, stingrays, and sharks sense the Earth's magnetic field, which helps them on their long-distance migrations.[10] The same is true of dolphins and whales, and it has been suggested that the beaching of these animals might sometimes happen because of disturbances in the magnetic field.

---

[10] Keller et al. [8].

It has been demonstrated that European robins and many other birds, including pigeons, rely on the Earth's magnetic field to navigate during migrations. Some birds seem to perceive the magnetic inclination to assess the latitude; others use the declination and follow a certain angle to the magnetic north when migrating. The polarity of the geomagnetic field guides bats to navigate for many hundreds to thousands of kilometres.

Marine turtles return to the same breeding beach to deposit their eggs after travelling through the ocean for thousands of kilometres. They make return travel with no visual landmarks, but only by following the Earth's magnetic field.

Magnetic storms from the Sun, generated by solar flares and coronal mass ejections, carry billions of tons of charged plasma particles with their magnetic fields. These strike the Earth's magnetosphere, causing strong disturbances. In addition to being a threat for astronauts and the Earth satellite system, communication and electric power grids, solar storms also offset the animals' magnetic sense of direction. It has been demonstrated that the nocturnal navigation capability of European robin is reduced in response to these phenomena.

The mechanisms of **magnetoreception**—the ability to detect magnetic fields—remain unknown. An increasing body of evidence suggests that a leading role in orienting living organisms is played by the presence within their bodies of tiny crystals of magnetite, a few nanometers in size (1 nm = one billionth meter = 10^{-9} m); these nanocrystals are used as a sort of compass to detect the surrounding magnetic forces. However, many animals do not contain these materials inside their body; their magnetoreception is more complex and involves various receptors whose biochemical activity is affected by the ambient magnetic field.

Given the relevance of the geomagnetic field for the living organisms, the question arises about the response animals might have during reversals. In these cases, adverse consequences could derive from both the polarity modification that upsets the orientation system and the intensity reduction by about 90% of the geomagnetic field strength, which dramatically reduces the protective action against solar radiation.

According to some geologists, mass extinctions attributable to magnetic reversals might have happened at least a couple of times shortly before the Phanerozoic and during the Mesozoic. The first case could have occurred some 550 million years ago, when a series of fast Earth's magnetic field reversals took place. Models, supported by some paleomagnetic data, suggest that the weakened Earth's dipole reduced the efficiency of the magnetosphere, allowing incoming cosmic radiation to reach the surface of the Earth and enter the shallow marine environment. The arrival of this increased radiation is thought to have triggered a widespread extinction at the end of the Ediacaran period, causing various soft organisms living in shallow marine waters to disappear. Another extinction related to magnetic reversal may have occurred between Triassic and Jurassic, when reduced protection by the magnetic shield resulted in the destruction of molecular oxygen by solar radiations, causing the onset of global anoxic conditions that were responsible for the extinction of various species.

However, many geobiologists consider that a causal relationship between magnetic reversals and mass extinctions lacks convincing evidence, and therefore are duly sceptical of these ideas. Their view is that animals are able to escape catastrophic consequences during reversals because they use various senses to migrate to suitable habitat. Moreover, reversals take several hundred to a few thousand years to complete. This leaves organisms much time to adapt to the new conditions. Finally, the attenuation of magnetic intensity is not long enough to significantly modify atmospheric composition.

References

1. Witze A. (2019) Earth's magnetic field is older than scientists thought. Nature 576:347
2. Hulot G, Balogh A, Christensen UR, Constable C, Mandea M, Olsen N (2010) The Earth's magnetic field in the space age: an introduction to terrestrial magnetism. Space Sci Rev 155:1–7
3. Sagnotti L, Scardia G, Giaccio B, Liddicoat JC, Nomade S, Renne PR, Sprain CJ (2014) Extremely rapid directional change during Matuyama-Brunhes geomagnetic polarity reversal. Geophys J Int 199:1110–1124
4. Creer KM, Irving E, Runcorn SK (1954) The direction of the geomagnetic field in remote epochs in Great Britain. J Geomagn Geoelectr 250:164–168
5. Runcorn SK (1956) Palaeomagnetic comparisons between Europe and North America. Proc Geol Assoc Canada 8:77–85
6. Heirtzler JR, Le Pichon X, Baron JG (1966) Magnetic anomalies over the Reykjanes Ridge. Deep-Sea Res 13:427–433
7. Stern DP (2002) A millennium of geomagnetism. Rev Geophys 40:1–30
8. Keller BA, Putman NF, Grubbs RD, Portnoy DS, Murphy TP (2021) Map-like use of Earth's magnetic field in sharks. Curr Biol 31:1–6

5

Seismicity—The Breath of a Restless Earth

E la possanza / qui con giusta misura / anco estimar potrà dell'uman seme / cui la dura nutrice, ov'ei men teme / con lieve moto in un momento annulla / in parte, e può con moti / poco men lievi ancor subitamente / annichilare in tutto. (They can also estimate/with a just measure/the power of humankind,/which the harsh nurse, when least expected,/with a slight motion in a moment annuls/a part, and can with a motion/barely less slight, in an instant/annihilate utterly. Wild Broom, or The Desert Flower, by Giacomo Leopardi, translated by Steven J. Willett)
Giacomo Leopardi, La Ginestra, o fiore del deserto (1836)

5.1   Introduction

The Earth is a restless planet, subjected to continuous changes in all parts of its body. We all know of winds blowing, water flowing, tides fluctuating, and other movements in the fluid Earth through our daily experiences. Movements of the Earth's lithosphere are subtler, but they do occur continuously and are manifested by seismicity, sometimes in a violent and dreadful way.

The earthquakes that shake the Earth's surface are extremely frequent. They take place continuously, both on the seafloor and the mainland, with more than one million events happening each year. Fortunately, most of them are feeble and detected only by seismographs; others are stronger, but their frequency is low and drops sharply with increasing energy.

Anyone can follow the Earth's seismic activity through various websites, including that of the US Geological Service (USGS) and the Berkeley Seismology Laboratory, where the location and energy of events are displayed shortly after their occurrence by coloured spots on a map.[1] Thus, modern technology allows us to momentarily observe the internal pulsations of our planet, which has been happening for billions of years and will continue, in the same way, indefinitely in the future.

5.2 Rock Failure, Earthquakes, and Faults

An earthquake is the sudden shaking of the ground that results from the seismic waves generated by rock failure somewhere within the Earth's crust. To understand how this happens, it is necessary to consider some technicalities of rock mechanics, simplifying the issue as much as possible.

5.2.1 Stress and Rock Deformation

The rocks beneath the Earth's surface are subject to the load pressure of the overlying rocks. This **lithostatic stress** acts with the same intensity in all directions and, therefore, does not cause rock failure.

However, in some regions of the Earth, large-scale tectonic forces generate additional stress (**differential** or **deviatoric stress**) that has an unequal intensity in different directions. This type of stress is superimposed on the normal lithostatic pressure and causes rocks to change their shape (**rock deformation**). There are three types of differential stresses: tensile, compressional and shear stress, which respectively yield a stretching, squeezing, and shearing (slippage parallel to stress) of rocks. The measure of the change of shape of rocks under the action of stress is referred to as **strain**.

Rocks (or any other material) respond in various ways to differential stress. If the stress is low, rocks slightly deform, but then revert to their initial shape once the stress is removed. In this case, the deformation is not permanent, and the rock is said to exhibit **elastic** behaviour.

If the stress is higher than the cohesive strength of the rock, deformation becomes permanent.

Such a state can be achieved either by a fracture of the rocks or by continuous modification of their shape, depending on rock compositions and ambient temperature and pressure. When a rock breaks, the failure is called

---

[1] https://earthquake.usgs.gov/earthquakes/map/; https://www.mapbox.com/bites/00267/.

brittle, and the rock is said to have **rigid** mechanical characteristics. If permanent deformation does not produce a fracture, deformation is referred to as **ductile**; in such a case rocks flow in the manner of viscous fluids, and their behaviour is indicated as **plastic**. Ductile deformation is accomplished by continuous and permanent adjustments of its constituent mineral granules.

Rocks exhibit variable mechanical behaviours. Those making up the lower continental crust and the mantle react plastically to deviatoric stress, thus modifying their shape without fracturing. By contrast, rocks of the upper continental crust and the oceanic crust behave elastically when subjected to low-intensity stress (**elastic deformation**). However, when the stress exceeds the threshold, the rocks break (**brittle failure**), releasing the accumulated energy as vibrations, i.e. seismic waves.

In conclusion, rocks of the upper continental crust and the oceanic crust behave as rigid bodies and can break to generate earthquakes; rocks of the lower continental crust and the mantle are softer and more malleable, thereby undergoing progressive plastic deformation rather than fracturing. These regions of the Earth, therefore, are (or should be) aseismic, i.e. free from earthquakes.

5.2.2 Earthquakes and Faults

The **elastic rebound theory** provides an explanation of how seismic energy is released during earthquakes. The idea was first suggested by the American geologist Harry Fielding Reid (1859–1944) after the great San Francisco earthquake in 1906, and is still valid today. According to this theory, rigid rocks affected by deviatoric stress first undergo elastic deformation and accumulate energy; the deformation goes on slowly until the rock strength is exceeded, leading to a fracture that releases the accumulated elastic energy as seismic waves. An earthquake, therefore, is the consequence of the sudden release of energy that has accumulated slowly over the years, centuries, or millennia of elastic deformation (Fig. 5.1).

The process of rock failure and earthquake generation can be visualised as a twig that is forcibly bent between arms: initially, the wood deforms elastically, i.e. it returns to the initial shape if the stress ceases. If the stress continues, then the strain accumulates until the twig breaks, producing noticeable vibrations. In this simple example, the twig represents rocks in the Earth and the vibration, the earthquake.

The place inside the Earth where the rupture of rocks, and the earthquake, starts is called **hypocentre** or **focus**. The **epicentre** is the point on the earth's surface directly above the hypocentre.

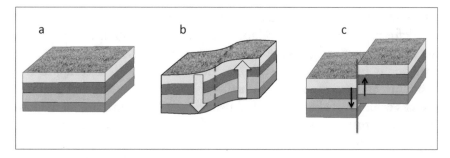

Fig. 5.1 Elastic rebound model of earthquake generation for rocks affected by deviatoric stress. **a** Initial undisturbed rocks; **b** Accumulation of elastic deformation due to action of shear forces (large yellow arrows): the rocks do not break and elastic deformation accumulates in the rock body (dashed line); **c** Fracturing and elastic rebound (spring back) of deformed rock layers (small black arrows): the accumulated elastic energy is released suddenly during fracturing as seismic waves during rock rebound along a fracture plane

Earthquakes rarely occur as single events; they often gather in clusters of different energy (foreshocks, mainshocks, and aftershocks). Such behaviour appears obvious when one considers that rock failure in the lithosphere occurs in several stages and a fracture can make nearby ones reactivate when in a precarious stability state.

The fracture along which the rock bodies are dislocated is the **fault**. The fault plane is the physical surface that separates two adjacent sliding rock blocks (Fig. 5.2a). In general, the longer the fault dislocation, the more violent the earthquake.

The type of fault varies depending on the orientation of the stress acting on the rocks. A **direct** or **normal fault** is generated by tensile (diverging) stress, a **reverse fault** is caused by compressional stress, and a **strike-slip fault** is a near-vertical fracture produced by shear stress (Fig. 5.3). In some cases, inclined reverse faults, called **thrust faults,** are generated between rock blocks that shift over one another.

Faults are the local effects of large-scale 'tectonic' movements of the crust, or the entire lithosphere. In short, blocks of the lithospheric, or tectonic **plates,** move continuously across the Earth's surface at a rate of a few centimetres per year (Chap. 6). Stress and strain build-up in particular areas, mostly along the edges of the plates. Here, rock failure occurs preferentially, making plate boundaries strongly faulted and highly seismic zones. Some earthquakes can be linked to local phenomena, such as magma movements under active volcanoes, infiltration of water from the surface or fluids from depth along

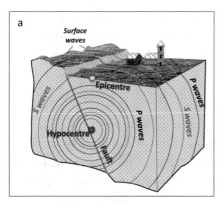
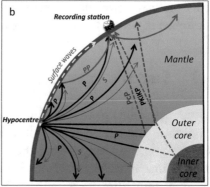

Fig. 5.2 **a** Schematic representation of an earthquake along a fault, with the production of P-waves (black lines) and S-waves (red lines) that spread across the Earth in all directions (body waves). The interaction between the body waves and the Earth's surface generates surface waves (yellow dashed lines); **b** Simplified paths of 'direct' P- and S-waves, and some reflected and refracted waves. Each wave is given a different name, depending on its travel path. For instance, PP-wave is an original P-wave that has been reflected by the Earth's surface; PcP-wave is a P-wave that is reflected by the outer core

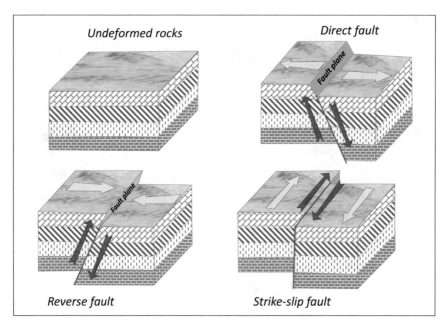

Fig. 5.3 Direct, reverse, and strike-slip faults. Yellow arrows indicate the orientation of tectonic stress acting on the rocks. Red arrows highlight the relative movements of the rock blocks along the fault plane

active faults, or even anthropic activities, such as the extraction of water and oil, forced injection of fluids into subsurface rocks, or nuclear explosions.[2]

5.3   Seismic Waves

Brittle rock failure generates vibrations or **seismic waves** that originate at the hypocentre and spread out in all directions through the body of the Earth. Wave propagation results from the elastic oscillation of constituent particles in rocks that move slightly from their positions, push the adjacent particles, and then return to the initial positions. Therefore, seismic waves transport energy, but not matter.

In some waves, called **longitudinal** or **compressional waves**, particle oscillations are parallel to the direction of propagation, and the energy is transferred by continuous compressions and relaxations of the particles in the same fashion as the coils move in a helical slinky spring toy (Fig. 5.4a).

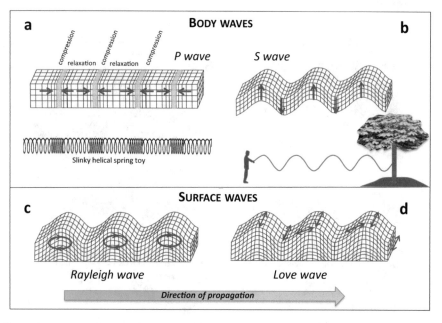

Fig. 5.4 Propagation mechanisms of P, S and surface waves. The small red arrows indicate the movements of the particles inside the rocks during seismic wave propagation. The large arrow is the direction of seismic waves propagation

---

[2] Seismic monitoring is the most widely used method for controlling nuclear explosions. These generate different waveforms than natural earthquakes and are therefore easily recognisable.

Transverse or **shear waves** travel by perpendicular particles' oscillations to the direction of wave propagation, a movement similar to that of a stretched rope whose end is shaken vertically (Fig. 5.4b). Compressional waves propagate both through solid and liquid materials; transversal waves only travel through solid bodies and cannot cross liquids.

Compressional waves are the fastest, reaching the recording stations first during an earthquake; for this reason, they are called **Primary Waves** (*Undae Primae,* **P**). The transverse waves are slower and, therefore, are called **Secondary Waves** (*Undae Secundae,* **S**). Obviously, the interval between the arrival times of P- and S-waves at a recording station becomes longer with the increasing distance from the hypocentre. The P- and S-waves are called **body waves** because they propagate through the body of the Earth. They are the only ones to be generated at the earthquake hypocentres.

A different type of seismic waves, called **surface waves**, are formed when body waves cross physical discontinuities of the planet. The **Rayleigh** (R) and **Love** (L) waves are generated by the interference of the body waves with the Earth's surface. They propagate along the surface, but oscillations also affect the Earth's interior to a considerable depth. The propagation mechanisms of surface waves involve both longitudinal and transverse motions of particles inside the rocks (Fig. 5.4c, d).[3] Therefore, they are felt as both up-down and side-to-side ground motions during earthquakes. Further, surface waves have greater amplitudes than body waves and cause the most vigorous shaking and damage.

The velocity of P-waves (V_P) ranges from less than one kilometre per second in loose sediments to 2–4 km/s in common sedimentary rocks and 4–6 km/s in igneous and metamorphic rocks. S-wave velocity (V_S) is about one half to two-thirds of V_P. Surface waves are about 10% slower than S-waves.

Seismic waves follow numerous paths when crossing the Earth. During an earthquake, P-waves and S-waves originating at the hypocentre propagate in all directions through the body of the Earth. They typically show curved trajectories, as a consequence of continuous variations of the mechanical characteristics of rocks with depth. Some waves reach the surface directly; others penetrate deep into the Earth, undergo reflection and refraction by various discontinuities, and bounce from one surface to another, changing their energy and propagation mechanisms.[4] As a result, several types of waves reach the surface at different times after an earthquake. A very incomplete representation of seismic wave paths is shown in Fig. 5.2b. A more realistic

---

[3] See website https://www.youtube.com/watch?v=B-8H4GmSNO8.
[4] Storchak et al. [1].

idea can be gained by consulting dedicated sources; for instance, the website of the International Seismological Centre.[5]

5.4   Seismographs and Seismograms

A **seismograph** is an instrument that records the ground motion generated by earthquakes, volcanic eruptions, or explosions. In its most basic (and easy to understand) configuration, it consists of a rotating drum covered with paper, and a weight (**inertial mass**) hanging from a spring or attached to a pendulum (Fig. 5.5). The end of a recording pen attached to the mass is placed onto the drum cover and leaves a mark on it. The pendulum or spring and the drum are mounted onto a frame that is firmly attached to the ground. When the ground shakes, the entire device moves, except for the mass and the attached pen, whose movements are delayed by inertia. Therefore, the pen draws a curve on the rotating paper that reproduces the oscillations of the roll, i.e. the motion of the ground.

Models designed to record horizontal oscillations have the rotating drum in a horizontal position and the inertial mass suspended to a pendulum (Fig. 5.5a). Vertical oscillation recording devices contain the drum in an upright position and the inertial mass and the pen suspended to a spring (Fig. 5.5b).

Modern seismographs are an upgrade to the instruments described above. There are no more papers and pens; the ground motion with respect to the

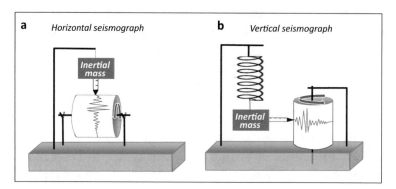

Fig. 5.5 Sketch of two seismographs designed for recording horizontal and vertical ground oscillations

---

[5] http://www.isc.ac.uk/standards/phases/.

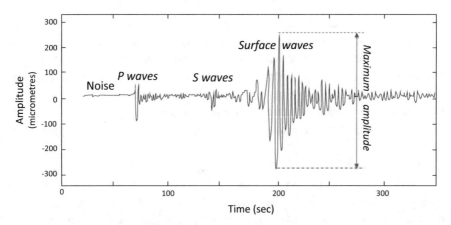

Fig. 5.6 An example of seismogram recording the arrival of P-waves, S-waves, and surface waves. The arrival times of seismic waves are shown on the horizontal axis, while the amplitude of ground oscillations is reported on the vertical axis. The time interval between arrivals of various wave types widens with the distance of the recording station from the hypocentre

mass is measured electronically. Modern instruments are extremely sensitive and capable of detecting sub-millimetre ground displacements.

A **seismogram** is the recording of ground shaking, classically reported on paper and presently on modern digital systems. Seismograms typically appear as irregularly oscillating lines generated by the various types of waves reaching the surface (Fig. 5.6). Arrival time in seconds is reported on the horizontal axis, whereas the ground displacement is on the vertical axis.

The seismograms contain numerous other oscillations, in addition to those generated by P-waves, S-waves, and surface waves. These complexities are linked to the arrival of the high number of waves that reach the seismograph after being reflected, refracted, and modified by Earth's discontinuities. Arrivals of waves generated by a single earthquake can last anytime from seconds to a few days, but the strong motion that causes damage is much shorter, ranging from seconds to around a minute or, more rarely, to a few minutes.

5.5 Earthquake Magnitude, Intensity and Frequency

Seismograms allow calculation of the **magnitude** of earthquakes, a parameter introduced in 1935 by Charles Richter (1900–1985). The **Richter magnitude** of an earthquake is defined as the logarithm of the maximum amplitude measured on the seismogram, corrected for the distance from the epicentre and other local factors. The 'zero magnitude' is defined as causing a horizontal oscillation of the ground by one-thousandth of a millimetre on a standard type of seismograph at 100 km from the hypocentre. Lower magnitudes can be measured by modern seismographs and have negative values. Because of the intrinsic character of the logarithmic scale, each step of one unit on the Richter scale represents a tenfold difference in the magnitude of an earthquake.

The Richter scale has many shortcomings, especially when earthquakes are very powerful or occur too far from the recording station. Therefore, seismologists developed other scales, such as the **moment magnitude** (M_W) and the **surface wave magnitude** (M_S) scales.[6] The Richter scale has been revised and used to measure local earthquakes (**local magnitude scale** M_L).

Like any measure, even the most accurate, magnitude is affected by an error, which is around 0.2–0.3 in the most optimistic estimates.

The magnitude of an earthquake depends on the type and size of the fault and the extent of rock displacement along the fault plane. Dislocations of a few centimetres produce moderate-energy earthquakes, while there are movements of meters or decametres along large faults for the most powerful earthquakes. For example, the Sumatra–Andaman earthquake of December 26, 2004, which had a very large magnitude of $M_W \sim 9.3$, was accompanied by a displacement of rock blocks of about 20 m along a fault 1200–1300 km long. The maximum values of magnitude are measured for large movements occurring along reverse or strike-slip faults and overthrusts. Therefore, the most powerful earthquakes occur in those areas that are affected by compressional and strike-slip stress.

Magnitude does not directly record the amount of energy released at the seismic source. However, studies by Richter and Gutenberg established an empirical relationship between the two parameters. To use a well-known term of comparison, an earthquake with $M_W \sim 6.0$–6.5 releases an amount of energy equal to that of the Hiroshima atomic explosion (about 15,000 tons of TNT, equal to 60×10^{12} J). Each increase in a unit of magnitude results in an increase in energy by about 31 times. For example, the energy released by the Messina (Italy) earthquake of 1908 ($M \sim 7.2$) was equivalent to approximately 500,000 tons of TNT. The Valdivia (Chile) earthquake of May 22,

---

[6] All magnitude scales retain the logarithmic character and are constructed in such a way to show similar values for the intermediate magnitude events. For very powerful earthquakes, M_w has significantly higher values than M_L. Erroneously, most media still refer to all magnitude scales as Richter magnitudes.

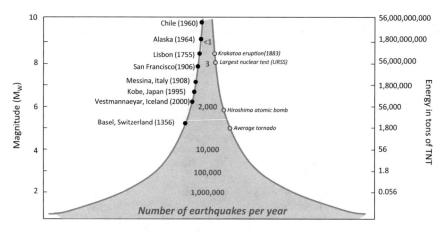

Fig. 5.7 Number of earthquakes (red numbers) with various magnitude and energy occurring yearly on Earth. Some natural phenomena and man-made events are shown by the yellow points

1960, the strongest earthquake ever recorded ($M_W \sim 9.5$), released an amount of energy of several billions tons of TNT.[7]

Scales of **earthquake intensity** measure the destructive effects at a given location rather than the earthquake strength. The most famous intensity scale is the one devised by the Italian volcanologist and geophysicist Giuseppe Mercalli (1850–1914), successively modified by Adolfo Cancani (1856–1904) and August Sieberg (1875–1945). It consists of twelve degrees, indicated in Roman numerals, with intensity increasing from the first degree (an earthquake felt only by instruments) to the twelfth degree (destruction of buildings, objects thrown into the air, etc.).

There is a rough relationship between the Mercalli scale intensity (I_0) and Richter magnitude (M_L), significantly changing with local geological factors. For most places in the Italian peninsula, a Mercalli scale intensity of the 'VI degree' approximates a local magnitude $M_L \sim 4.3$, and the intensity of 'XI degree' corresponds roughly to $M_L \sim 7$. Such a relationship, although inaccurate, is useful to estimate the magnitude of historical earthquakes for which there are no instrumental records.

The number of earthquakes occurring both at the local and planetary scales is inversely proportional to their strength (Fig. 5.7; see also Table 5.1). Events with $M_W \sim 9.0$ occur worldwide once every 20 years, on average. As the magnitude decreases, the frequency decreases from 4 per year for earthquakes

---

[7] The SI unit of energy is 1 J (J). A joule is the energy needed to lift by one meter an object, weighing about 100 g. The energy of nuclear explosions is often given in megatons (Mt), equivalent to 1,000,000 tons of TNT. A megaton of TNT equals about $4.2 \& x 10^{15}$ J.

Table 5.1 Some of the strongest earthquakes of the last century

Locality	Date	Magnitude	Locality	Date	Magnitude
Colombia-Ecuador	31 January 1906	8.4	Kuril Islands, Russia	13 October 1963	8.5
Messina, Italy	28 December 1908	7.1	Alaska, USA	4 February 1965	8.7
Haiyuan, China	16 December 1920	8.3	Irpinia, Italy	23 November 1980	7.0
Atacama, Chile	11 November 1922	8.3	Izmit, Turkey	17 August 1999	7.6
Kamchatka, Russia	3 February 1923	8.4	Sumatra, Indonesia	26 December 2004	9.3
Indonesia	1 February 1938	8.5	Sumatra, Indonesia	28 March 2005	8.6
Assam, China	15 August 1950	8.6	Kashmir, India	8 October 2005	7.6
Kamchatka, Russia	4 November 1952	8.9	Jakarta, Indonesia	12 September 2007	8.5
Dodecanese, Greece	7 September 1956	7.8	Santiago, Chile	27 February 2010	8.8
Alaska, USA	9 March 1957	8.6	Tohoku-Oki, Japan	11 March 2011	9.1
Valdivia, Chile	22 May 1960	9.6	Sumatra, Indonesia	11 April 2012	8.6
Alaska, USA	28 March 1961	9.3	Nepal	25 April 2015	7.8

with M_W ~ 7–8, to one per day for M_W ~ 4–5, and up to a thousand per day for M_W ~ 1–2. People can feel only earthquakes with magnitudes higher than 1.5–2.0.

5.6   Global Distribution of Earthquakes

Our Earth is characterised by a very uneven regional distribution of seismicity (Fig. 5.8). Some large areas, both on the continents and the ocean

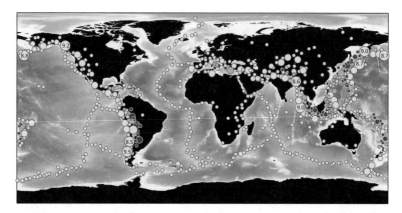

Fig. 5.8 Distribution of seismicity on Earth. The most powerful (wider circles) and deepest (green and blue colours) earthquakes occur along the margins of the Pacific Ocean, Indonesia, the Lesser Antilles, the South Sandwich Islands, and the Himalayan-Alpine chain. Numbers in the circles indicate the magnitudes (M_W) of some of the strongest earthquakes (see Table 5.1). The topographic base of the map is taken from the NASA website

floors, show little or no seismicity (e.g. Australia, Siberia, Greenland, Antarctica, Central Pacific Ocean). In contrast, other regions are sites of a very large number of events. Zones of particularly dense seismic activity include the Pacific Ocean margins, the Sunda Archipelago, the Lesser Antilles, the Alpine-Himalayan belt, the mid-ocean ridges, the East Africa Rift Valley, the Basin and Range region of the western USA, and a few other areas. Notably, active volcanism also concentrates in most of these zones (Chap. 3). However, there is no causal relationship between the two phenomena, since only a small fraction of earthquakes originate from volcanic activity. In reality, seismicity and volcanism are both linked to the same processes, namely the extensional tectonics along the oceanic ridges and East Africa, and the compressional stress associated with lithospheric convergence in the circum-Pacific margins, the Caribbean, the Sunda Islands and the Alpine-Himalayan belt (Chap. 6).

Figure 5.8 shows that earthquakes with the highest magnitudes, indicated by larger circles, are concentrated along the margins of the Pacific Ocean and in the Indonesian Archipelago, where about 80% of the Earth's total seismic energy is released; intense seismicity also occurs in the Antilles Archipelago and along the Alpine-Himalayan orogenic belt (about 15% of total seismic energy). These zones are dominated by compressional and/or shear-stress tectonic regimes due to lithospheric plate convergence or continental collisions (Chap. 6). Seismicity is weaker and less frequent along mid-ocean ridges and the rift valleys where lithospheric faulting is prevalently related to extensional stress.

Most earthquakes are shallow (yellow circles in Fig. 5.8), with hypocentres extending to the maximum depths of a few tens of kilometres. Shallow foci are quite obvious, given that earthquakes can only occur by the failure of rigid rocks, restricted to the crust. However, in the compressional stress zones, intermediate and deep earthquakes are also detected (green and blue circles in Fig. 5.8), with hypocentres sometimes extended to a depth of 600–700 km.

A crucial aspect of deep seismicity consists in the distribution of earthquake foci that are not disseminated chaotically but are orderly distributed along planes that cut through the upper mantle with different angles. These planes are named **Wadati-Benioff seismic planes** (or simply **Benioff zones**) in honour of the geophysicists Kiyoo Wadati (1902–1995) and Hugo Benioff (1899–1968), who first discovered the geometry of deep seismicity under Japan and the Tonga-Kermadec Islands. As an example, Fig. 5.9 schematically illustrates sections of Benioff zones beneath a limited sector of South America and Japan. These are just two cases of deep seismic planes that, although segmented by large transversal faults, characterise the entire circum-Pacific belt and occur worldwide in the other zones of deep seismicity such as Indonesia, the Lesser Antilles, South Sandwich and the Mediterranean area.

Deep seismicity is a sort of geological conundrum, since it is located within the upper mantle whose rocks should deform plastically. Thus, the occurrence of Benioff planes can only be explained by assuming that rigid crustal rocks occur at depth, embedded inside plastic material.

The recognition of Benioff zones is one of the main discoveries of modern geology. As discussed in Chap. 4, the geomagnetic investigation of the ocean floor during the middle of the last century led to the conclusion that mid-ocean ridges are zones of extensional tectonics where new oceanic crust and lithosphere are continuously created. Recall that such a hypothesis raised much scepticism, as it required that either the planet was expanding or that the lithosphere was consumed somewhere else on Earth.

The discovery of Benioff zones provided the solution to these problems, leading to the recognition that the lithosphere forming along the ocean ridges migrates away on both sides of the ridge axis and then is transported into the Earth's mantle along the Benioff-Wadati planes. The distribution of earthquake foci along inclined planes is a piece of substantial evidence for the occurrence of crustal slabs residing inside the mantle. Seismicity is related to the fracturing of rigid crustal rocks; the planar distribution of earthquake foci reflects the flat geometry of the slab.

The transport of the lithosphere into the mantle is referred to as **subduction** (from the Latin word *subdùcere*: remove, take away), emphasising the

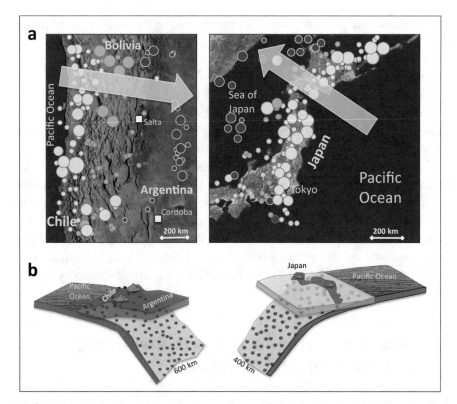

Fig. 5.9 **a** Distribution of seismicity in regions of the Andes and Japan; colours of circles indicate the depths of the hypocentres, ranging from less than 50 km (yellow) to 50–200 km (green) and more than 200 km (blue). The size of circles denotes magnitudes; **b** Geological sections across two transects of South America and Japan (large arrows in the upper panels) showing the distribution of earthquake foci (red dots)

concept that in these geological settings, the lithosphere is removed from the surface and brought into the mantle.

5.7 Summary

Earthquakes are the most evident manifestation of the vitality of Earth. They are generated inside the crust by the brittle failure of rigid rocks. Fractures can be induced by tensile, compressional and shear stress that gives direct, reverse, and strike-slip faults, respectively.

The Earth's seismicity is concentrated along the edges of the Pacific Ocean, Indonesia, the Caribbean, South Sandwich and the Himalayan-Alpine belt, where the largest magnitude earthquakes occur and most of the global seismic energy is released. Some seismicity also occurs along the mid-ocean ridges and in continental rifts such as in East Africa, Rhyne graben (Germany), and the Basin and Range (Western USA). Almost all seismic zones are associated with active volcanism.

Most earthquakes occur within the crust and at a depth of a few kilometres. However, there are long belts characterised by both shallow and deep earthquakes, whose foci extend along inclined planes to the depths of several hundred kilometres. Deep-focus earthquakes are distributed along flat surfaces that are referred to as **Benioff-Wadati planes**. Belts of deep focus seismicity occur around the borders of the Pacific Ocean, the Lesser Antilles, the South Sandwich Islands, and in some sectors of the Mediterranean area.

The Wadati-Benioff planes highlight the occurrence of large blocks of rigid oceanic crust and lithosphere within the mantle. These are called subduction zones and are the places where the lithosphere, which is formed at the ocean ridges, sinks deep into the Earth and is ultimately destroyed by melting within the mantle.

5.8   Box 5.1- Earthquake Effects, Prediction, Forecasting, and Mitigation

The collapse of buildings, bridges, and other structures are the destructive effects of earthquakes. The level of damage depends on several factors, including the intensity and duration of shaking, the regional and local geology, and the construction quality of buildings. The heaviest death toll of an earthquake occurs in zones where building quality is inadequate to support shaking, such as in old historical towns and villages, a particularly severe problem in southern Europe and large regions of Asia.

Local geological factors also have a role in the destructive power of earthquakes. For instance, shaking at sites on soft sediments is more extensive and lasts for longer than at locations of hard rocks. Therefore, the use of appropriate construction techniques and careful anti-seismic renovation of old buildings effectively limit damages and casualties during earthquakes. One of the strategies to protect old structures is fastening the walls and floors of buildings together. Such a procedure makes up a sort of single stiff structure that then behaves coherently during the shaking.

The collapse of buildings and bridges is not the only negative consequence of earthquakes. They can also cause soil liquefaction and trigger landslides, avalanches, and tsunamis—all of which may cause enormous destruction and loss of life.

Landslides and **debris flows** in unstable steep hillsides and mountains are commonly triggered by earthquake shaking. Once activated, landslides and debris flows travel downhill, traversing roads and rivers and sometimes moving at high speed along valleys, sweeping away anything on their way.

Liquefaction is a phenomenon that affects water-saturated loose sediments and causes sedimentary material to behave like a liquid. Usually, the granules of sand are in contact with each other and contain water in their pore spaces. When shaken by an earthquake, the grains lose coherence and temporarily transform the solid sediments into a viscous liquid-like material with minimal bearing strength. Buildings constructed on liquefying soils experience a loss of support, which results in tilting, structural damage, and even collapse. In some cases, the expulsion of water during shaking results in an eruption of wet sand.

Tsunami (from the Japanese term *tsu nami*, meaning port wave) is a dreadful effect of submarine earthquakes. A strong-magnitude earthquake (M > 6–7) occurring at shallow depth below the seafloor or lake bottoms can quickly lift or lower the rocks, displacing huge masses of water. This water rapidly moves away from the source area as rolling waves, spreading in all directions. The wave movement affects the entire water column. The height of waves in the open sea is modest, whereas the speed is several hundreds of kilometres per hour. As the tsunami approaches the coast and the seafloor becomes shallower, the speed of waves reduces and the height increases, sometimes reaching tens of meters. Waves lose very little energy as they travel and can produce catastrophic consequences, even in areas located thousands of kilometres away from the source. The tsunami triggered by the 26 December 2004 earthquake off the coast of Sumatra not only wrought destruction locally but also caused damage in coastal areas as far away as India and East Africa.

Like all waves, tsunami waves have crests and troughs. When the wave crest reaches the coast, the sea level rapidly rises like a quickly surging tide. In most cases, the trough reaches the coast first, drawing water away from the shore and exposing the seafloor. This withdrawal is followed by the crest, with a large wave crashing ashore. The rapid retreat of seawater from shoreline to an abnormal distance offshore is one of the first signs of a tsunami. Therefore, anyone on a beach who witnesses an unusual sea retreat would do well to move away quickly from the shore to a high place and stay there for hours. Tsunami is not a single wave and strikes with a series of waves that follow each other after minutes or even hours, and the first wave is not necessarily the most destructive.

Tsunamis are also caused by other processes such as underwater volcanic eruptions or landslides. Water displacement is activated by the collapse of volcanic structure, or the emission of large-volume pyroclastic flows or landslides that enter the water and displace equal amounts of liquid.

The tsunami that hit Sumatra during the eruption of Krakatoa in 1870 was triggered by huge pyroclastic flows entering the sea, an event that was measured on tidal gages as far away as the English Channel. A more recent and much less destructive tsunami occurred around noon on 30 December 2002 in the Southern Tyrrhenian Sea (Italy), due to a landslide detached from the western flank of the Island of Stromboli, a very steep and unstable active volcano in the Aeolian Islands. Several meters high, the waves reached the other islands of the archipelago in a few minutes and hit the coast of Southern Italy in less than one hour. The tsunami caused damage to the structures along the coastline and only avoided catastrophic consequences because beaches that are generally crowded from spring to fall were deserted during the winter.

Their abrupt onset and devastating effects make earthquakes some of the most frightening natural phenomena. For these reasons, incredible efforts have been devoted to understanding when an earthquake might occur. Two main approaches are used. The first is aimed at finding precursory phenomena at the local scale, in order to facilitate the short-term (days or months) **prediction** of an imminent earthquake in a given location. The other is directed at providing long-term **forecasting** by estimating the probability of an earthquake with a given magnitude occurring within a specific window of time in a given area.

Short-term prediction is useful only if it is able to provide reasonably accurate indications of the location, timing, and magnitude of an earthquake. Generic statements that say an earthquake will occur without specifying the time and place are risible, if not harmful. Nevertheless, they are guaranteed to be true, since earthquakes occur continuously worldwide.

Attempts to make reliable short-term predictions are based on the observation of several potentially interesting precursors such as local ground deformation, clusters of low energy foreshocks, variations in the chemical composition of groundwater, fluctuations in the local electromagnetic field, and changes in animal behaviour. One or more of these diagnostic phenomena have often been reported shortly before an earthquake. However, there are few, if any, examples of the successful prediction of imminent earthquakes based on these precursors.

A frequently cited, but unique case of successful near-event prediction is that of the earthquake that occurred in Haicheng, northeast China, on 4

February 1975. Chinese geologists observed changes in groundwater levels, the unexpected emergence of snakes from dens, and their deaths along the streets because of low temperatures, and several foreshocks. Together, these coincident occurrences convinced them that a catastrophic earthquake was imminent. The city of Haicheng was evacuated before it was struck by an earthquake of $M_S = 7.3$. However, the catastrophic earthquake that occurred in the following year at Tangshan ($M_S = 7.6$), some 400 km away from Haicheng, was not accompanied by any precursor; therefore, no precautionary measures were taken, and several hundred thousand people were killed or injured. In other cases, warnings released by scientists and civil authorities based on precursors have not been followed by an earthquake, although people had left their homes and slept in tents for weeks or months, as a precaution for possible impending danger.

To better understand the difficulties of earthquake prediction, even in areas particularly suited to these purposes, consider the case of Parkfield, California (USA). In this sector of the San Andreas Fault between San Francisco and Los Angeles, earthquakes had occurred at relatively regular intervals in 1857, 1881, 1901, 1922, 1934, and 1966. Since a new event was expected between 1988 and 1993, a large variety of instruments were installed by the United States Geological Survey (USGS), and many international teams arrived in the region with the objective of observing what would happen before the event. The appointment, however, was not kept; the next earthquake didn't occur until 11 years later, on 28 September 2004, apparently with little or no antecedent phenomena.

This and other similar experiences have led many geophysicists to conclude that there is a very low possibility of making reliable short-term earthquake predictions. Therefore, research has primarily focused on long-term forecasting to establish the probability of an earthquake with a given maximum magnitude occurring in a particular region. The result of this approach is the preparation of maps that highlight areas with different levels of seismic hazard. Since the ground acceleration during earthquakes is a significant seismic phenomenon responsible for building collapse, many hazard maps report values of maximum Peak Ground Acceleration (PGA) expected for earthquakes that may occur over a specific period. These maps are based on data furnished by global seismic monitoring networks, local seismic and geodetic monitoring networks, geological fieldwork, and records of historical earthquakes.

As an example, the seismic hazard map for the USA is shown in Fig. 5.10. Colours denote areas with different seismic hazards in ascending order from

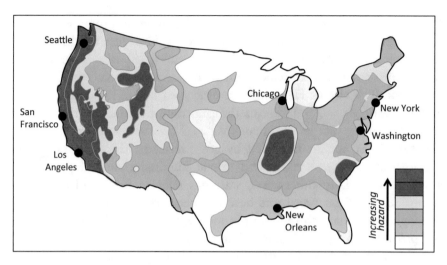

Fig. 5.10 Long-term seismic hazard map of the USA. Seismic hazard increases from white and blue to brown and red coloured zones. Simplified after the USGS seismic hazard map (https://www.usgs.gov/media/images/2018-long-term-national-seismic-hazard-map.)

white to red. Long-term forecasting and hazard maps can facilitate appropriate measures in building construction, service organisation, and civil protection systems (Table 5.1).

Reference

1. Storchak DA, Schweitzer J, Bormann P (2003) The IASPEI Standard Seismic Phase list. Seismol Res Lett 74:761–772

6

Plate Tectonics—The Great Unifying Theory

A scientific model is a good model if it is elegant, contains a few arbitrary or adjustable elements, agrees with and explains all existing observations, makes detailed prediction about future observations that can prove or falsify the model if they are not borne out.
Stephen William Hawking (1942–2018)

6.1   Introduction

The word **tectonics** (from the Greek τεκτονικός, tectonicòs = pertaining to building) is used in geology to indicate both the large-scale movements of the Earth crust and the branch of geology that studies these phenomena. Tectonics investigates the vertical and horizontal movements of the lithosphere that leads to mountain building, the opening and closing of the oceans, and other such first-order geological processes. The objective of tectonic theories has been to create a single and consistent framework that explains all of these geological processes, a goal finally scored by the theory of plate tectonics.

Plate tectonics is one of the most outstanding achievements in the field of Earth Sciences. It has the same significance for the Earth Sciences that the theories of evolution by natural selection and relativity have in the fields of Biology and Physics. Once synthesized and articulated in the early 1960s, plate tectonics developed rapidly to become the dominant scientific paradigm

© The Author(s), under exclusive license to Springer Nature
Switzerland AG 2021, corrected publication 2022
A. Peccerillo, *Air, Water, Earth, Fire*,
https://doi.org/10.1007/978-3-030-78013-5_6

in Earth Sciences, after only a short period of objections and opposition—a rapid, complete and universal success.

Plate tectonics is an outstanding scientific theory because it fits into a single coherent model many distinct geological phenomena which in the past appeared erratic and hard to correlate, such as the distribution and nature of earthquakes and volcanism, the building of mountain belts and oceans, the distributions of geological formations and fossils across the world, and more. Additionally, it possesses the prerequisites of every great scientific theory: it is simple, elegant and easy to understand, even for those unfamiliar with geology. A considerable number of scientists, mainly American and British, contributed to the construction of this theory. Most of the data used to develop the theory came from studies of the ocean floor. Oceanographic investigations began during World War II and expanded greatly afterwards during the 1950s and 1960s. The objectives of these studies were essentially military; however, a wealth of fundamental information about the composition and structure of the ocean floor were provided for geologists, greatly expanding their understanding of large unexplored regions of the planet and placing the foundation for the greatest cultural revolution in the field of Earth Sciences.

The Earth is the only planet of the solar system where plate tectonics is at work; a main stream of research in astrophysics is aimed at investigating whether it might be active in some of the newly discovered planets outside the solar system.[1] The Earth contains liquid water on the surface and is home to complex life. Such a concurrence is not accidental because, as it will be discussed later in this and the next chapters, the occurrence of liquid water makes plate tectonics possible. Plate tectonics preserves the dynamic equilibrium among the various terrestrial environments, thus creating favourable conditions for the birth and evolution of life. Therefore, liquid water, plate tectonics and life are closely related phenomena.

The plate tectonics paradigm germinated from an earlier theory known as **continental drift,** an audacious concept anticipated in the 1800s by various scientists, and systematically formulated during the early 1900s by the German meteorologist Alfred Wegener (1880–1930), a giant whose shadow still casts over modern Geology.

---

[1] Noack and Breuer [1].

6.2 Fixism *Versus* Mobilism

The idea that the continents have been immobile since their formation, with little if any lateral displacement, dominated Earth Sciences until the beginning of the last century. Such a theory is known as **fixism**.[2] The Austrian geologist, Eduard Suess, one of the main promoters of the fixist theory, argued that the Earth had gone through an initial period of near-total melting, a hypothesis also accepted by modern geologists. During this time, continents were built up by aggregation of aluminium silicates (**sial**) formed inside the liquid Earth and, due to their low density, accumulated by buoyancy on the surface of a magma ocean. Solidification of these light materials constructed the continents that preserved their positions and original dimensions until the present times. The only relevant movements were believed to be vertical, as demonstrated by the fossil remains of marine organisms observed high up in mountain chains worldwide. According to this view, the formation of mountain ranges resulted from thermal contraction and shrinking of the sialic crust (**contractionism**) during progressive cooling of the planet: a phenomenon similar to that observed on the skin of a desiccated apple.

Fixism and contractionism left many unresolved problems, including the distribution of Earth's mountain chains, arranged in long continuous belts instead of being uniformly distributed as ripples across the surface of the planet. Moreover, the rift valleys and ocean basins require an extension of the crust, which is at odds with the idea of a general contraction of the Earth. Therefore, **mobilist** theories were conceived which held that the continents are subjected both to uplift and lateral drifting, although movements are exceedingly slow to detect.

6.2.1 The Theory of Continental Drift

Continental drift was the first mobilist theory that challenged the fixist paradigm, about one hundred years ago. This idea was posited by Alfred Wegener in a seminal paper published in a German scientific journal and later extensively and thoroughly discussed in various editions of the book, *Die Entstehung der Kontinente und Ozeane* (The Origin of the Continents and Oceans).

The German scientist argued that the continental masses are not fixed, but rather continuously subjected to horizontal movements that cause them

---

[2] The same term is also used in biology to indicate theories suggesting that the presently living species are the same as in the past, and evolution has not occurred.

to slowly move over long distances. He considered the continents as huge rafts floating on a dense, warm and deformable substrate formed of magnesium silicates (**sima**). In their horizontal movement, the continents opened their way through the ocean floor, deforming and breaking the rocks apart, in the same way icebergs do when travelling across the seawater. The continents existing today were viewed as portions of a single vast supercontinent called **Pangaea** that dismembered many millions of years ago, leaving various fragments adrift.

The concept of continental drift was inspired by the excellent fit between the margins of some continents, particularly South America and West Africa, which were evocative of a formation by splitting apart of a single original larger block. Francis Bacon (1561–1626) and many others had already noticed the particular shape of these continental margins, as soon as the first fairly accurate topographic maps of Africa and South America became available. In the middle of the nineteenth century, more than half a century before Wegener, the Italian-French-American geographer Antonio Sneider-Pellegrini (1802–1885) considered the similarities of continental margin geometry and the distribution of fossils and geological rock formations on the two sides of the Atlantic to conclude that these Earth landmasses were once united in a single vast supercontinent and that South America had split off from Africa to migrate to its present position.

Evidence favouring a geological and palaeontological continuity between South America and Africa became overwhelming at the turn of the nineteenth and twentieth centuries, thanks to studies of many scholars, especially the South African geologist Alexander Du Toit (1878–1948). The followers of the fixist theories recognised the validity of these findings. Yet, they interpreted them as evidence for the existence of **geological bridges** connecting the two continents in the past; faunas would migrate through these bridges and spread out in both Africa and South America. An isthmus, or even a small continent, was postulated to have joined Africa and South America and subsequently sagged below sea level, like the mythical Atlantis. These imaginative hypotheses soon had to succumb to the greater scientific rigour of continental drift. According to continental drift theory, mountain ranges were built up along the leading edges of moving continents, as the rocks were compressed and deformed by resistance to their drift. In contrast, archipelagos such as the Antilles and the circum-Pacific island arcs were the product of detachment of **sial** fragments left behind at the tail of drifting continents. The vertical uplift of the crust was positively acknowledged, but its extent was considered much smaller than the thousands of kilometres horizontal travels of continents.

Wegener considered the continents as giant rigid bodies, while the oceanic crust through which they travelled was deformable. Such an assumption was one of the main flaws of the theory, since it was clear that the oceanic crust was made up of rigid rocks that could not part and open the way for continents. Another serious problem related to the cause of the continental drift, which remained unknown and led the famous British geophysicist Sir Harold Jeffreys (1891–1989) and many other contemporaries to reject the theory because, in their view, there was no force strong enough to move continents. The formation mechanism of mountain belts was also deemed unrealistic because an allegedly deformable oceanic crust could not twist and fold rigid continental rocks. For these and several other problems, the continental drift hypothesis was marginalised and had to wait for the mid-twentieth century to be revitalised, thanks to the palaeomagnetic studies summarised in Chap. 4.

6.2.2   From Continental Drift to Plate Tectonics

The studies of the ocean floor carried out in the 1950-1960s were crucial for the birth of plate tectonics. It was observed that the overall flat surface of the abyssal plains was interrupted by two giant structures, the submarine volcanic chain of the **mid-ocean ridges** and the **oceanic trenches**. The latter is a long and deep linear depression that develops for tens of thousands of kilometres along the Pacific Ocean margins, the border of the Indonesian-Banda island arc, the Lesser Antilles and the South Sandwich Islands, and, to a much minor extent, in the Mediterranean. It is typically 100 km wide and can have depths exceeding 10,000 m, if not filled with sediments. At the very beginning of the 1960s, Harry Hess (1906–1969) of Princeton University concluded that the mid-ocean ridges and trenches are the surface expression of convective movements within the Earth's mantle. He postulated that the oceanic ridges develop in zones of uprising hot convection currents, thereby explaining abundant basaltic volcanism, whereas the trenches are the places where the cooler branches of convective cells descend back into the mantle.

Further studies demonstrated that the boundary between the crust and the upper mantle (**Moho**) lies at a depth of about 8 km beneath most of the ocean floor, but rises to near the surface along the axial zones of the oceanic ridges. Deep-sea dredging and drilling found that the ocean floor is basically made up of a superimposed sequence of basaltic lavas, gabbro intrusions and, at greater depth, peridotite. This entire rock suite is affected by variable degrees of alteration by interaction with seawater at elevated temperature (**seafloor metamorphism**) and is covered by a layer of fine-grained deep-sea sediments whose thickness progressively decreases from the abyssal plains to the central zones of the ridges.[3] It soon became apparent that the thickness of the sedimentary cover, the symmetrical distribution of the ages and magnetic polarities on the seafloor, and the very shallow Moho along the ridge axis, all denoted that the oceanic basins are young geological structures that had been generated by seafloor spreading and formation of new crust along the oceanic ridges (Chap. 4).

Thanks to the work of Harry Hess and John Dewey in the US, Tuzo Wilson in Canada, Jason Morgan and Dan McKenzie in the UK, and Xavier Le Pichon in France, among several others, the body of petrological, geophysical and geomorphological evidence was finally incorporated into the general unifying theory of **plate tectonics**.

According to this theory, the oceanic lithosphere is created at the oceanic ridges, and successively displaced horizontally as large slabs or plates towards the trenches, where it sinks and disappears into the Earth's mantle. In this view, global Earth dynamics consists of a simple process of continuous creation, lateral displacement, and the ultimate foundering of lithospheric plates: a cyclical mechanism that requires no mass or volume modification of the planet.

A decisive innovation of plate tectonics with respect to Wegener's theory is that horizontal movements affect the entire lithospheric slabs or plates, not the continents alone. Continents are merely riding lithospheric plates and float with them over the underlying asthenosphere (see Chap. 1, Fig. 1.5). Wegener's theory was thus rehabilitated and significantly modified, overcoming many of the objections raised by the followers of fixism.

6.3   Plate Tectonics: The Framework

The theory of plate tectonics suggests that the Earth's lithosphere is divided into several large slabs or plates separated by deep faults. Plates have the shape of irregular rock caps that together comprise a giant jigsaw puzzle, enveloping the solid Earth (Fig. 6.1). Plates are mobile and jostle slowly but continuously across the surface of the Earth. The thickness of plates ranges from about 100 to some 250 km, and increases from the oceans to the continents, reaching a maximum under old stable continental regions (**cratons**). Surface extension

---

[3] Interestingly, similar rock sequences (sediments plus altered basalt lavas, gabbro, and peridotite) are found in many mountain chains, such as the Alps and the Dinaric-Hellenic mountain belt, and these are known as **ophiolites**. The ophiolite sequences represent fragments of the ancient ocean floor that was emplaced onto continents by tectonic movements, providing further proof of horizontal mobility of the Earth's crust.

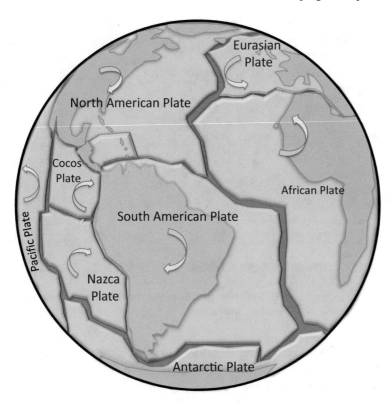

Fig. 6.1 Lithospheric plates enveloping the Earth. The arrows indicate rotation movements of various plates. The plate margins are tightly interlocked as pieces of a jigsaw puzzle, but are here represented as gaps to show the underlying mantle

of single plates spans hundreds of thousands to several tens of millions of square kilometres. The largest plates are shown in Fig. 6.2.

All the plates are made up of a rigid basal layer of peridotite (or **mantle lid**), overlain by either oceanic or continental crust. Some plates contain only oceanic crust; others have both continental and oceanic crust, passing laterally one to the other (see Chap. 1, Fig. 1.5). The continents are not individual blocks freely cruising through the Earth's surface, as hypothesised by Wegener; instead, they sit on the lid and, therefore, move because the entire lithosphere plate is in motion. The Pacific, Nazca, and Cocos plates are composed entirely of oceanic lithosphere, whereas North America, South America, Eurasia, Africa, and other plates consist of both oceanic and continental lithosphere passing sideways one to the other.

As recalled elsewhere in this book, the continental crust consists of relatively low-density rocks, such as granites, metamorphic, and sedimentary rocks; by contrast, the oceanic crust is mainly made up of high-density basalts

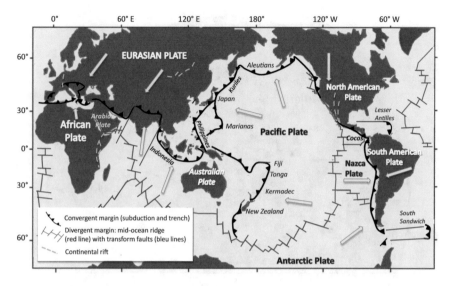

Fig. 6.2 Distribution and boundaries of the major Earth's lithospheric plates. The arrows indicate the relative plate movements

and gabbros. These contrasting densities impart a different buoyancy to the lithosphere, depending on the type of crust overlying the lid. The continental lithosphere—composed of the lid and continental crust—is relatively light and, therefore, decidedly 'floats' above the underlying asthenosphere to form almost all the Earth's landmasses. By contrast, the oceanic lithosphere is heavy and flattens onto the underlying mantle, forming the ocean basins, the large topographical depressions submerged by water. In short, the topography of the ocean floor and continental masses is a direct consequence of differences in lithospheric density.

The lithospheric plates behave like giant rigid slabs that travel horizontally, each in a particular direction, at a speed of a few centimetres per year. Since they lie on a spherical surface, their movement is rotational, as expected from Euler's fixed-point theorem (Fig. 6.1); therefore, each point of a plate describes arcs of circumference around a rotation pole.

Deep fractures zones mark the boundaries between adjacent plates and allow them to move independently. Along these boundaries, plates can move away from each other, collide or slide sideways to each other, depending on whether their relative movements are divergent, convergent, or tangential.

The **intra-plate** geological settings—i.e. the regions internal to the plates—are relatively stable and geologically quiescent. Almost all geological activity occurs at plate boundaries, where volcanism, seismicity, formation of the ocean basins, the rise of mountain chains, and many other geological processes take place.

The lithospheric plates grow continuously along **oceanic ridges** because of the addition of basaltic rocks and the cooling and stiffening of ascending asthenosphere, whereas they are destroyed by sinking into the mantle along converging margins—a process referred to as **lithospheric subduction**. The amount of lithosphere being created at the mid-ocean ridges is balanced by the lithosphere foundering into the upper mantle at subduction zones. Therefore, plate tectonics is a cyclical process that incessantly creates and destroys crust and lithosphere, providing a continuous exchange of matter between the internal and external regions of the Earth. Lithospheric plates have changed in number, size and shape with time. Currently, there are 14 larger and 38 smaller plates.[4]

The lithospheric plates move laterally on the Earth's surface, sliding over the underlying **asthenosphere**. As stated in the previous chapters, the asthenosphere is a mantle layer containing tiny amounts of magmatic liquid that makes the peridotite particularly ductile. The presence of the asthenosphere is a key factor for plate tectonics. It separates the lithosphere from the underlying mantle, allowing the plates to move laterally. Without the asthenosphere, the lithosphere would be firmly bound to the underlying mantle and could not break into mobile plates. In other words, the Earth would be a tectonically static planet, similar to Mars.

6.4   The Plate Boundaries

The boundaries between plates are fundamental geological sites. Therefore, it is important to take a closer look at what happens along them. As mentioned earlier, plate boundaries can be divergent, convergent and transform, depending on the relative movements of adjacent plates (Fig. 6.3). Relative plate motion can result from several absolute movements; for instance, convergence can occur either because two plates move against each other or because one stands still and the other approaches or, finally, when both move in the same direction but at a different speed, which makes the faster plate push the slower one.

---

[4] Bird [2].

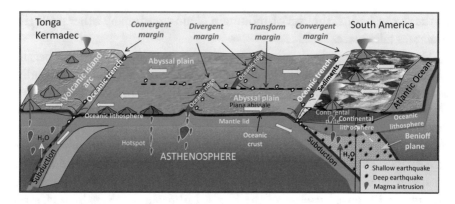

Fig. 6.3 Simplified geological section across the southern Pacific Ocean and the South American continent, from the Tonga-Kermadec islands to the western Atlantic Ocean. The oceanic lithospheric plates created at the mid-ocean ridge (divergent margin) slide laterally in opposite directions and converge with other plates at the western and eastern margins of the Pacific Ocean. Along convergent margins, the oceanic lithosphere descends into the upper mantle, causing seismicity and volcanism. In the western Pacific, the almost vertical subduction of ocean lithosphere under an oceanic plate margin generates volcanoes that form the Tonga-Kermadec island arc. The mid-ocean ridge in the central Pacific is cut by transform faults that separate two plates, sliding horizontally with respect to each other (transform margin). The yellow arrows indicate the relative movement of plates

6.4.1 Divergent Boundaries

The divergent boundaries are the places where two adjacent plates separate from each other, spreading out in opposite directions like two enormous conveyor belts. The spreading rate ranges from 1–2 cm/year (e.g. North Atlantic ridge) to about 15–16 cm/year (e.g. East Pacific ridge).

Plate divergence results in extensional faulting and lithospheric thinning, which causes the ascent and partial melting of the to produce huge amounts of basaltic magmas. Magmas intrude along faults, erupt on the surface and solidify to form new crust; contemporaneously, cooling and stiffening of the ascending asthenosphere transform the uprising peridotite into a rigid lid. Together, the basaltic rocks of the crust and the underlying peridotite of the lid create new lithosphere. For this reason, divergent plate margins are commonly referred to as **constructive boundaries**. As new mantle material and basaltic magmas are added, the previously formed lithosphere is pushed in opposite directions away from the divergent margin, providing additional space to create new crust and lid.

Diverging margins cut across both the ocean floor and the continents. In the oceanic regions, the diverging plate margins build up the mid-ocean ridges. These have an elevation of a few thousand metres and emerges above the sea surface in Iceland (the northern branch of the mid-Atlantic ridge) where there is a high rate of magma production, because of the occurrence at depth of a thermally anomalous mantle region called **hotspot**. The mid-ocean ridges contain elongated rift valleys bordered by normal faults in their axial zones and are cut by perpendicular strike-slip (transform) faults that divide the ridges into segments (Fig. 6.3). The intense fracturing promotes the infiltration of seawater deep into the hot volcanic crust and its return to the ocean surface as **hydrothermal fluids**. This hydrothermal activity contributes substantially to the dispersion of the Earth's internal heat and provides an effective chemical exchange between ocean water and rocks of the oceanic crust.

The oceanic lithosphere moving away from oceanic ridges uniformly flattens on the underlying asthenosphere, forming the **abyssal plains**. Many **seamounts** of volcanic origin rise from the ocean floor, sometimes emerging above the surface to form islands such as Galapagos, Hawaii, Maldives, St. Helena, Azores, and the Canary Islands. These volcanoes are sometimes arranged in lineaments that can extend over a length of several thousand kilometres (Box 6.1).

In continental areas, this extension produces alignments of tectonic valleys delimited by normal faults (**rift valleys**), which are sites of shallow seismicity and volcanism. If continued over time, divergent movements can lead to the widening of rift valleys, finally resulting in an ocean basin (Chap. 4, Fig. 4.6). The **Great African Rift** of East Africa is a major zone of current continental divergence; it extends from Mozambique to Afar, the Red Sea, and the Gulf of Aden. This rift separates the Somali and Arab plates to the east, from the African continental plate to the west. The Red Sea, the Afar triangle, and the Gulf of Aden are in a more advanced stage of extension than the other parts of the Great African Rift and have now reached the stage of an embryonic ocean basin. In the next million years, the Great Rift Valley will stretch to form an ocean basin.

Overall, divergent margins are sites of extensional tectonic regime, ascent and decompression melting of the asthenospheric mantle, spreading, formation of abundant basaltic magmatism, generation of new crust and lithosphere, and the opening of ocean basins.

6.4.2 Convergent Boundaries

Convergent boundaries are the contact sites of two converging plates. Along these margins, the plate with higher density slides down under the margin of the other (**overriding plate**), first descending and then disappearing into the mantle along **subduction zones** (Fig. 6.3). For this reason, convergent margins are also referred to as **destructive boundaries**. There is a substantial balance between the mass of the lithosphere created at oceanic ridges and that consumed along destructive margins. It has been estimated that about 3.0–3.5 km^2 of crust is consumed every year along subduction zones, and a similar amount is created at the constructive margins.[5]

The subduction zones do not form a single continuous system, but are divided into sectors separated by transcurrent faults; all sectors have a similar structure, but each of them differs somewhat from the other by geometry, seismicity, and volcanism. A typical example is the subduction zone of the Aleutian Arc in the north-western Pacific. Another more extreme case is the Italian peninsula along which several distinct subduction sectors of the Adriatic plate and Ionian seafloor under the Apennines have been recognised over a distance of a few hundred kilometres.[6]

The tectonic style along converging margins is dominated by compressional and shear stress, which results in rock folding and reverse, thrust, and strike-slip faulting. However, a few hundred kilometres away from the plate contact, the overriding plate is subjected to extensional tectonic stress, which favours magma ascent and volcanism. In some cases, especially in zones of west-dipping subduction such as the Toga-Kermadec Islands (Fig. 6.3) and Japan, strong extensional tectonics results in the opening of oceanic-type basins behind volcanic arcs (**backarc basin**).

Subduction is only possible for the oceanic lithosphere, which has a sufficiently high density to sink into the upper mantle; by contrast, the continental lithosphere cannot founder because of its low density. Consequently, the oceanic crust is a temporary geological structure, which is continuously created along the ridges and consumed by subduction. This contrasts starkly with the continents, which are stable and, once formed, can be preserved for billions of years, although worn down by exogenic geologic processes.

As mentioned in Chap. 5, subduction causes earthquake clusters whose hypocentres lie along flat planes (**Benioff-Wadati zones**) that dip at variable angles beneath the margin of the overriding plate, away from oceanic trenches (Fig. 6.3). Seismic activity at Benioff-Wadati zones is possible because the subducting crust has rigid mechanical characteristics and undergoes fracturing that releases mechanical energy. Deep earthquakes occur nowhere else on Earth because there are no other places where rigid rocks occur inside the mantle. The increase in pressure and temperature to which the lithosphere

---

[5] Rowley [3].

[6] Peccerillo [4].

is subjected during subduction causes the transformation of oceanic basalts into metamorphic rocks, especially **eclogite**, whose high density (about 3.5–3.7 g/cm^3) favours plate sinking into the peridotitic upper mantle (density ~ 3.3–3.4 g/cm^3).

Once the lithospheric slab reaches the transition zone in the mantle at about 650 km depth, it flattens and splits into multiple pieces that disperse through the upper mantle or continue their descent until reaching the core-mantle boundary. Here they accumulate to build up the **D" layer** (Chap. 1), which apparently is the deepest level reached by crustal material. According to many scholars, the D" layer can become gravitationally unstable and release very hot plastic rocks that rise as mantle blobs or **plumes**. Once they reach the base of the lithosphere, plumes undergo decompression melting to generate intra-plate volcanism.

Oceanic trenches are the obvious result of downward bending of the oceanic plate beneath the overriding margin, and are a main topographic indicator of converging plate boundaries. They are zones where sediments, either scraped off the subducting oceanic plate surface by the overriding plate margin or being transported from an adjacent continent, are accumulated, sometimes forming thick sequences (**accretionary prism**) that contribute to lateral growth of continents (Box 6.2). Volcanic material (e.g. seamounts) can also be scraped off the ocean floor and become part of the accretionary prisms.

Magmatism is generated by hydrous melting of the upper mantle above the subduction zones (Chap. 3); abundant water released by the oceanic crust triggers melting and is also the reason for the high explosivity of the volcanoes that form above subduction zones. Magma compositions in subduction environments vary widely from basalt to rhyolite, with the prevalence of andesite. Magmas erupt at the surface along fronts of volcanic edifices situated on the margin of the overriding plate, some 100–150 km away from the oceanic trench. The emplacement of these magmas contributes significantly to the growth of continents.

The so-called **Circum-Pacific Ring of Fire** is by far the best-developed chain of volcanoes along converging plate margins (Fig. 6.3 and Chap. 3, Fig. 3.4); other examples are present in Indonesia (e.g. Krakatoa, Merapi, Tambora), the Lesser Antilles (e.g. Soufrière, Montserrat), South Sandwich (e.g. Mount Belinda, Bristol Island) and the Mediterranean Sea (e.g. Stromboli, Vulcano, Santorini, Nysiros).

Metamorphism is another important petrogenetic process of converging plate margins. It affects the rocks when these are subjected to high pressure and temperature (Chap. 1, Box 1.1). Metamorphism, therefore, takes place both along the subduction zones and within the crust of the over-riding plate, beneath the volcanic chains where high heat flow is released by uprising magmas. Metamorphic transformation affecting the oceanic crust and lid along subduction zones generates high-density rocks, such as **eclogite**, which facilitates the plate foundering into the mantle.

Orogenesis (or **orogeny**) is the most colossal process occurring along converging plate boundaries. Some mountain belts, or **orogens**, develop because of the thrust exerted by the undergoing oceanic plate on the over-riding margin; this type of mountain chain is referred to as **non-collisional orogen** (Fig. 6.4a). In other cases, deformation, crustal thickening and uplift result from the collision between two continents, with one riding up over the margin of the other (**collisional orogen**; Fig. 6.4b). Both types of orogeny involve deformation and thickening of the crust that is pushed upwards to form mountain chains and downwards to form a kind of orogenic root, extending down into the mantle.

The Andes are an example of a non-collisional mountain range; they have been formed because of the compression of the Nazca and Antarctic

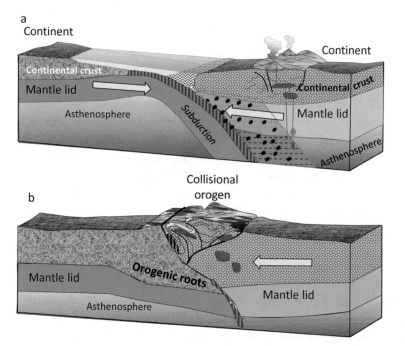

Fig. 6.4 **a** Non-collisional orogen (Andean type) in which mountain uplift occurs because of the thrust of the oceanic plate against the continental margin; **b** Collisional orogen (Alpine-Himalayan type) formed by the collision between two continents, after the interposed oceanic crust has been consumed by subduction.

oceanic plates against the Pacific margin of South America. By contrast, the Himalayan-Alpine chain is a collisional orogen formed along the impact belt of the African, Arabian, and Indian continents against the Eurasia.

Many other orogens have been formed during the geological past (e.g. the Appalachian Mountains, Scottish Highlands, Bohemian Massif, and the Ural Mountains, among others). These areas have lower altitudes and more gentle relief than younger orogens because of hundreds of million years of erosion by water and ice.

Not all elevated areas, commonly called mountains, are linked to orogenesis[7]. Apart from volcanoes, there are extensive regions of the Earth, such as the Ethiopian and Colorado plateaux, that have been uplifted by the upper mantle ascent, perhaps under the influence of large upwelling plumes. However, there is no crustal thickening and intense and pervasive folding in this situation, as occurs in orogenic belts.

To sum up, converging margins are the sites of compressional tectonics and subduction, which generate intense seismicity at variable depths, the formation of oceanic trenches, sedimentation, magmatism, and metamorphism. Importantly, it is at converging plate margins that mountain ranges form through processes referred to as **orogenic**.

6.4.3 Transform Boundaries

Transform boundaries are large faults, along which one lithospheric plate slides past its neighbour (Fig. 6.3). Transform boundaries are sites of shallow, sometimes strong, seismicity and sporadic volcanism, but no creation or destruction of the lithosphere occurs; therefore, they are termed **conservative margins**. A renowned transform boundary is the San Andreas Fault in California, which separates the North American and the Pacific plates, respectively moving in south-eastward and north-westward directions. Additional examples include the boundaries between the Cocos and the Nazca plates, the margins of the Caribbean plate, the Scotia plate in the southwestern Atlantic, and many others (Fig. 6.2).

---

[7] Most dictionaries define a mountain as a natural elevation of the Earth's surface having a considerable mass and altitude and deeply eroded valleys. From a geological perspective, a mountain range implies crustal deformation and thickening along a converging plate boundary.

6.5   Why Do Plates Move?

Forces that move plates have been, and still are, lively debated among geologists. It has long been hypothesised that thermal convection within the mantle, initially envisaged by the English geologist Arthur Holmes (1890–1965), might drag the plates, thus providing the ultimate engine for plate tectonics (Fig. 6.5a).

According to this model, internal movements are activated as the hotter lower mantle becomes gravitationally unstable and rises upward. By contrast, cooler and heavier mantle at the top sinks towards the core-mantle boundary, thus closing the convective cell. The rising mantle would comprise a large plume that impacts against the lithosphere, spits it apart and pushes the resulting plates in opposite directions—a **bottom-up tectonics** whose primary engine is deep in the Earth. Therefore, divergent margins would occur above the rising currents of the convective cells, while the converging margins are related to the descending currents of the convection cells, as suggested by Harry Hess. Besides these large movements, there would be minor plumes that, starting from the D" layer, rise towards the base of the lithosphere, producing intra-plate volcanism. Detailed tridimensional seismic images (**seismic tomography**) support a whole mantle convective stirring.

There are some objections to the model of whole mantle convection. First, there are doubts that the deep mantle rocks have a density low enough to activate convection, because the dilation effect due to the rise in temperature would be opposed and nullified by the increase in pressure. Moreover, seismic studies show that subduction zones become nearly horizontal when they reach depths of 650–670 km, which contrasts with the whole mantle convection hypothesis; on the other hand, subduction is sometimes traced down to the deepest lower mantle. Finally, a system of convective cells extending across

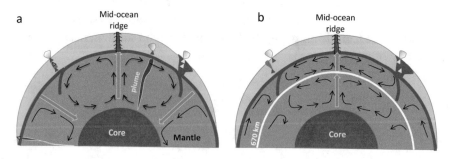

Fig. 6.5 **a** Model of whole mantle convection and plume ascent; **b** Two-layer convection model

the entire mantle are expected to produce somewhat equally spaced lithospheric breakup, with similar dimensions for all lithospheric plates. In reality, neighbouring plates sometimes have strikingly unequal sizes. For instance, the contiguous Cocos and Pacific oceanic plates have approximate areas of 2.9 million km^2 and 102 million km^2, respectively. These disparities would require the coexistence of adjacent convective cells with very contrasting dimensions. Such a conclusion does not necessarily conflict with whole convection model, but simply implies more chaotic systems that result in instabilities and irregularities in the mantle flow.

There is a general agreement, however, that whole mantle convection has been operating during the early stages of the Earth history, in the Late Hadean to Early Archaean, when the mantle was much hotter and fluid than at present. Whole mantle convection during this time could have induced fracturing of the primitive crust and initiation of a sort of embryonic plate tectonics regime.

Another hypothesis, accepted by most geologists, is that there are two distinct convective systems in the mantle, one operating within the lower mantle and a second extending vertically from the transition zone to the base of the lithosphere (Fig. 6.5b). Movements in the upper layer would be the effect, and not the cause, of plate mobility; in essence, the gravitational sinking of the lithospheric slab along the subduction zones would pull the entire plate (**slab pull**) and force the convective movements inside the upper mantle and the transition zone.[8] According to this model, the upper mantle would behave passively with regard to tectonic movements since the entire system would be governed by the gravitational sinking of the lithosphere along subduction zones (**top-down tectonics**).

Italian geologists Carlo Doglioni and Giuliano Panza postulate that the weight of the lithosphere alone is insufficient to determine plate mobility. The two authors have argued that an additional push to lithospheric plates could be provided by the tangential forces of lunar attraction.[9] This would also explain why the subduction zones dipping to the east (for example, under the Andes) are less inclined than those with westward orientation (for instance, under the Tonga-Kermadec island arc; see Fig. 6.3).

---

[8] Anderson [5].
[9] Doglioni and Panza [6].

6.6   Where, When and Why Does Subduction Start

A final question about plate tectonics is where and when the oceanic lithosphere undergoes large-scale fracturing and begins foundering into the mantle, to start subduction.

An essential condition for subduction to start is that the ocean lithosphere has to be affected by compressive tress to cause downward arching. It is also necessary that the lithosphere is sufficiently dense to sink into the underlying mantle. Since the oceanic lithospheric plates cool and become progressively denser while moving away from the mid-ocean ridges, high densities are reached only for the portions that are very old and generally situated near continents. The presence of physical discontinuities, such as ancient fracture zones or the physical change of rocks at the transition between oceans and continents, are additional factors that favour the lithosphere to break and subduction to start.

Another critical requirement is the surface liquid water that can penetrate through rock fractures, react with the minerals of the lithosphere (e.g. olivine), and transform them into hydrated phases (e.g. serpentine). Such mineral alteration produces a mechanical weakening of the lithosphere, making it easier to bend.[10]

Once subduction has started, water is transported into the mantle to ultimately return to the surface with volcanism. However, a certain amount of this water remains at depth to cause incipient partial melting within the asthenosphere, the soft layer of the mantle over which plates slide, thus making plate tectonics possible. In essence, water is essential both to trigger subduction and facilitate plate dynamics.

Subduction comes to an end when the oceanic lithosphere between two converging continental plates is wholly consumed and the continental masses collide. In some cases, new lithospheric fractures can develop within the plates after collision, and a new subduction process can start in a position parallel to the previous one. Such a process has occurred along several converging margins, such as in the western United States. For example, the western portions of Washington state (US) and western British Columbia (Canada) were not part of the North American plate until the Middle Cretaceous, but were successively appended to the continent by a subducting plate coming from the west. After these terrains were accreted, subduction jumped westward into its present position. A similar process affected the Balkan Peninsula

---

[10] Regenauer-Lieb et al. [7].

during the Cenozoic, when a series of terrains were accreted to Eurasia by the northward moving African plate, a process that was accompanied by southward jumps of the subduction zone until reaching its present position in the eastern Mediterranean Sea, offshore the Peloponnesus and Crete.

6.7 Summary

Plate tectonics is a revolutionary scientific theory that has the same significance for Geology as evolution by natural selection had for Biology and relativity had for Physics. Developed during the 1960s, it provides a simple but comprehensive framework for most geological process, past and present. Plate tectonics is exclusive to the Earth and is not observed in the other planets of our solar system.

According to plate tectonic theory, the Earth's lithosphere — formed by the rigid portion of the uppermost mantle (lid) and the overlying oceanic crust – is divided into multiple adjacent lithospheric plates of various sizes that float on the underlying mantle and move horizontally at a speed of centimetres per year. The asthenosphere is the soft layer of the upper mantle sitting beneath the lid that separates the lithospheric plates from the underlying mantle. Thanks to the asthenosphere, plates can move, jostling on the Earth's surface and making plate tectonics work.

The boundaries between adjacent plates comprise deep fracture zones along which plates move apart, converge, or slip past one another. Plate margins are the places where the most important geological processes occur.

Divergent margins separate adjacent plates that spread away from each other; they are characterised by extensional faulting, ascent and melting of the asthenosphere, and formation of vast quantities of basaltic magmas. These divergent margins may pass through continents and oceans, forming rift valleys and mid-ocean ridges, respectively. The mid-ocean ridge is a chain of submarine volcanoes some 65,000 km in length; it stretches across all the Earth's oceanic basins, rarely emerging to the surface.

Converging margins are the places where two adjacent plates move against each other. Here, a denser oceanic plate slips under the edge of a less dense oceanic or continental plate and sinks into the upper mantle. Such a process is known as subduction. Converging margins are characterised by deep oceanic trenches, intense seismic activity extending from a depth of a few km to about 650–700 km, metamorphism, magmatism, and formation of mountain ranges (orogenesis). Presently active subduction zones are located along the borders of the Pacific Ocean, the Sunda Islands, the South Sandwich, the

Lesser Antilles, and the central-eastern Mediterranean Sea. Finally, transform margins separate adjacent plates that slide sideways past each other. The San Andreas Fault in California is the best-known transform margin.

Plate tectonic theory envisages the Earth as a dynamic system in which there is continuous competition between global phenomena operating in opposing directions. Extension along the ocean ridges creates new lithosphere, which subsequently is destroyed in subduction zones. Horizontal plate mobility is promoted by the gravitational sinking of heavy oceanic lithosphere into the mantle. The weight of the subducting lithosphere drags the entire plate and is the very reason for plate mobility.

6.8   Box 6.1—Linear Volcanism and Hotspots

Volcanoes occurring inside the lithospheric plates, away from plate boundaries, are sometimes aligned along rows that run over considerable distances of up to thousands of kilometres. Many geologists interpret the linear distribution of volcanoes as an effect of thermal and/or chemically anomalous zones (**plume**) positioned at a fixed location in the mantle generating a **hotspot**. Such plumes would be rooted in the lowermost mantle and persist for tens of millions of years, continuously undergoing melting and delivering magmas to the surface. Magmas generated at these hotspots pierce through the lithosphere and construct linear alignments of volcanoes on the surface, as the lithospheric plate moves over the hotspot. The final result is a series of volcanoes whose orientation marks the direction of the plate movement.[11] The age of volcanism changes along the trend, becoming progressively older away from the hotspot.

The best-known example of such linear volcanism, which can be easily observed on Google Maps, is the Hawaii-Emperor chain, a series of some 130 volcanic islands and seamounts present on the Pacific Ocean floor from the Hawaiian Islands to western Aleutians, over a distance of 5,800 km (Fig. 6.6). Volcanism is active at the southeastern end of the chain on the island of Hawaii, which is presently located above the hotspot magma source. Volcanoes on the other Hawaiian Islands and the western and northern sectors of the Hawaiian-Emperor chains are no longer active and become progressively older approaching the western Aleutians, where the Detroit seamount has an age of about 80 million years. Interestingly, the Hawaii-Emperor chain

---

[11] Morgan [8]. However, some authors consider the plume hypothesis poorly supported by data, invoke a shallow origin for magmas and explain linear volcanism as a by-product of plate tectonics. See: Foulger and Natland [9] .

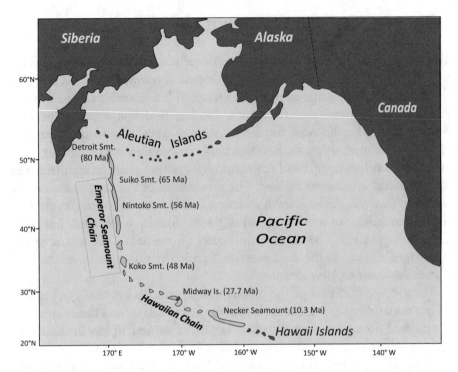

Fig. 6.6 Sketch map of the Hawaii-Emperor chain, running from the Hawaiian Islands to the western Aleutian Islands. Numbers indicate ages in millions of years. The hot spot is presently located beneath the southern extremity of the Hawaiian archipelago, where active volcanism is occurring and a submerged volcano (Loihi) is forming

has two segments with different orientations. The Koko seamount, about 1,000 km northwest of the Midway Islands, is the inflexion point where the volcanic alignment bends at an angle of about 60°. Volcanism in the Koko seamount area is about 48–50 million years old, denoting that there was a drastic modification of the motion of the Pacific Plate over the hotspot, or a shift of the source of magmatism. Some authors believe that such a drastic change in the motion of the Pacific Ocean might be a consequence of the initial stages of collision between the Indian and Asian continents, which formed the Himalayan Mountains. Such linear volcanism occurs at other places. Some examples are the island chain of French Polynesia in the southwestern Pacific, the Mascarene-Réunion chain in the southwestern Indian Ocean, the Cameroon Line in the Gulf of Guinea-Cameroon.

6.9   Box 6.2—The Continents

The geological meaning of the word *continent* is substantially different from that of common parlance and non-geological literature. Most dictionaries define continents as large continuous masses of lands, such as Asia, Europe, Africa, the Americas, and Antarctica. Geologists define a continent as a portion of the Earth's crust formed by various types of igneous, metamorphic, and sedimentary rocks, having a bulk composition matching that of an andesite. Continents can be of any size, are not restricted to the mainland and include some submerged regions. The dimensions of the continents can be immense (Eurasia, Africa, North America, South America, Australia, Antarctica), or medium to small (e.g. Madagascar, Arabia, Greenland, Sardinia-Corsica, Seychelles). The biggest underwater continent is Zealandia, a largely submerged region in the southwestern Pacific extending from offshore of eastern Australia to New Zealand.

The average thickness of the continental crust is 35 km that reaches 60–70 km under the great mountain ranges of the Himalayas and the Andes. The continental lithosphere can extend to depths of some 250 km in the oldest cratons, where geological stability for billions of years has favoured cooling of a large part of the uppermost mantle and its transformation into a rigid lid.

The continental crust is highly heterogeneous, comprising various types of igneous, metamorphic, and sedimentary rocks. However, there is a vertical compositional variation from the upper to the lower continental crust, sometimes marked by a second-order seismic discontinuity (**Conrad discontinuity**). The upper crust has an average composition similar to granite, dominated by quartz and alkali-feldspars. By contrast, the lower crust contains higher amounts of Fe–Mg minerals such as amphibole, pyroxene, biotite, and garnet. Such a variation is the effect of chemical differentiation processes that operate within the crust. In short, melting processes inside the lower crust generate granitic magmas that migrate upward, resulting in the differentiation of the continental crust into two compositionally distinct layers.

Overall, the continental crust has a density of about 2.5–2.6 g/cm^3, much lower than the underlying mantle. This density makes the continental crust 'float' on the mantle to form the subaerial landmasses. Its low density also inhibits the continents from being involved in subduction. However, some very dense portions of the lowermost continental crust can sometimes detach and sink vertically into the underlying mantle—a process referred to as **crustal delamination**.

The margins of continents can be described in a tectonic sense as being active or passive. **Active continental margins** are at the boundaries of plates and, therefore, are the sites of intense deformation, seismicity, and volcanism. A typical example of an active continental margin is the western border of South America, along which the Nazca and Antarctic oceanic plates are subducting into the upper mantle. **Passive continental margins** are stable and lack significant tectonic and volcanic activity. They are located inside the plates and are the sites where the continental crust passes laterally to oceanic-type crust. The Atlantic margins of South America and Africa are two examples of passive continental margins.

Continents are first-order geological structures whose formation has been the main legacy of Earth's history. Besides making up the land masses on which we live, they are the sources of the chemical elements that make ocean water 'salty'. Phosphorus, potassium, iron, silicon and other elements that are essential for marine life all originate from continents and are brought to the ocean by river and stream water. Continents drive the oceanic and atmospheric currents and play a key role in regulating the global climate.

The largest continents are geologically complex structures consisting of various large domains of rocks with different ages and compositions. These are the ultimate result of a series of geological events and processes that started billions of years ago and are still continuing today.

The birth of the first proto-continents is traced back to the Hadean when partial melting of the primordial basaltic crust produced bodies of granite magma. These ascended towards the surface and cooled to form tiny emerged islands in the immense proto-ocean of water that for some time enveloped the entire planet. Most of these blocks were subsequently destroyed by erosion and probably recycled into the mantle; but others survived and formed the seeds of future continents. The first two billion years of the Earth history were dominated by basaltic magmatism that created oceanic crust, and it has only been in the last two billion years the continental crust become volumetrically significant.[12]

The earliest continents started growing through various mechanisms. The partial melting of the basaltic crust and the addition of magmatic material from the mantle and the lower crust along the subduction zones were decisive for continental birth and growth. Accumulation of sediment in the accretionary prisms along converging margins gave a significant contribution to lateral continental growth.

Plate tectonics has played a crucial role in the evolution of continents. Plate convergence led to the collision of various continents and their aggregation into single large **supercontinents**. These were too big to survive for long and

---

[12] Taylor and McLennan [10].

were broken apart by the movements of the underlying mantle to form several separated smaller continents, which successively reaggregated to form a new supercontinent. Such a cyclical process, known as the **Wilson cycle**, occurred several times during the history of the Earth (Chap. 9). During each cycle, the mass of continents increased significantly by the addition of magmas from the mantle and lateral accretion of sediments.

The present continents are patchworks of interlocked terranes, showing different compositional, and structural features and ranging in age from Archaean to Cenozoic. Africa (Fig. 6.7) is a typical example of such a hetero-

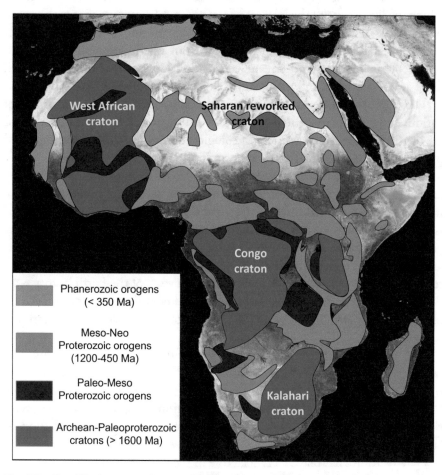

Fig. 6.7 Simplified structural map of Africa. The present continent consists of various domains that were formed independently at different times and were successively aggregated by plate tectonics

geneous structure. North America also has a similar composite nature, being an assemblage of about a dozen mini-continents formed between 3.5 billion years ago and the Cenozoic. This feature inspired the naming of the **United Plates of America** for North America.

References

1. Noack L, Breuer D (2014) Plate tectonics on rocky exoplanets: Influence of initial conditions and mantle rheology. Planet Space Sci 98:41–49
2. Bird P (2003) An updated digital model of plate boundaries. Geochem Geophys Geosyst 4:1027
3. Rowley DB (2002) Rate of plate creation and destruction: 180 Ma to present. Geol Soc Am Bull 14:927–933
4. Peccerillo A (2017) Cenozoic volcanism in the Tyrrhenian Sea region. Springer 399 p
5. Anderson DL (2001) Top-down tectonic? Science 293:2016–2018
6. Doglioni C, Panza G (2015) Polarised plate tectonics. Adv Geophys 56:1–167
7. Regenauer-Lieb K, Yuen DA, Brandlund J (2019) The initiation of subduction: criticality by addition of water. Science 294:578–580
8. Morgan WJ (1971) Convection plumes in the lower mantle. Nature 230:42–43
9. Foulger GR, Natland JH (2003) Is "Hotspot" volcanism a consequence of plate tectonics? Science 300:921–922
10. Taylor SR, McLennan SM (1995) The geochemical evolution of the continental crust. Rev Geophys 33:241–265

7

Geochemical Cycles—The Circulatory System of Planet Earth

Potrei raccontare di atomi di carbonio che si fanno colore o profumo nei fiori; di altri che, da alghe marine a piccoli crostacei, a pesci via via più grossi, ritornano anidride carbonica nelle acque del mare, in un perpetuo spaventoso girotondo di vita e di morte […] (I could tell about carbon atoms which become colour or perfume in the flowers; of others that, from seaweed to small crustaceans, to increasingly larger fish, return as carbon dioxide to the sea waters, in a perpetual frightening circle of life and death […].)
Primo Levi, Il sistema periodico (1975)

7.1 Introduction

Many geological processes are cyclical, and progress by a sequence of steps that repeat, in the same order, indefinitely.

Plate tectonics has an inherently cyclical nature. It basically consists of continuous creation and destruction of the lithosphere, and uninterrupted transfer of material from the mantle to the crust and then back to the mantle. Other similar processes include the lithogenic cycle, with its continuous transition from one rock type to the other (Box 7.1), and the sedimentary cycle with the recurrence of weathering, transport, deposition, and creation of new rocks (Chap. 2).

All these processes involve the circulation of matter throughout the various spheres of the Earth. The cyclical migrations of chemical elements through natural environments are known as **geochemical cycles**. We speak of **biogeo-chemical cycles** when the elements move through both geological and biological systems.

Geochemical cycles are major processes that affect the entire Earth system, regulating its composition and evolution. In a sense, they represent a sort of circulatory system for the planet, which, as in the human body, trans-fers substances, deposits waste, and regulates the overall equilibrium of the entire organism. The geochemical cycles have been active since the early stages of the Earth evolution; they have changed significantly through geological time, because of global planetary modifications such as climatic variations, the continuously growing size of continents, the variable distribution of continental masses and oceans, the abundance and distribution of living organisms, and many other factors, including human activities.

Knowledge about the behaviour of the chemical elements during their journeys through the Earth's spheres is pivotal to understanding the evolu-tion of the planet and how the Earth system works. Yet, such a key issue is often overlooked and finds little space in the popular scientific information and general geology books. Hence, it is appropriate to address this omission by summarising some basic aspects of geochemical cycles and their effect on the external environment: two topics that are the subject of the present and the next chapter.

7.2   Geochemical Cycles: Some Definitions

The behaviour of the chemical elements in the Earth system is governed by both the intrinsic characteristics of the individual elements (atomic number, oxidation state, ionic radius, etc.) and the chemical-physical conditions present within different geological domains. Therefore, each element follows its own geological path, even though the different cycles intersect and affect each other.

In the following pages, only some of the geochemical cycles that have a particular interest for the environment and the evolution of the Earth system will be considered. The water, sodium, and carbon cycles will be discussed in some detail, with brief summaries provided for cycles of phos-phorus, nitrogen, oxygen, iron, and a few other elements. For illustrative purposes, the geochemical cycles will be shown as particular diagrams called **box models** or **system models**. In these diagrams, rectangles represent the various natural environments or **geochemical reservoirs**, whereas arrows symbolise the direction of migration of matter from one environment to another, i.e. the **element flux** between reservoirs.

Numerical parameters will be sometimes considered to quantify the mass of reservoirs, the annual fluxes, and the time spent by individual elements in the various reservoirs (**residence time**). These numbers often represent rough estimates that can differ from one author to another and are subject to continuous revision, as new knowledge is gained. Nevertheless, such estimates are useful for a better evaluation of the role of geochemical cycles in the evolution of the planet and their impact on the equilibrium of the external environment.

7.3   The Water Cycle

Water is a primordial substance inherited from the solar nebula, which gave birth to our solar system more than four and a half billion years ago. Falling meteorites and comets have subsequently provided some additional minor amounts.

Although concentrated in the hydrosphere, water is found in the atmosphere, soil and rocks and, at a very low concentration, inside the mantle and perhaps even in the core. It is also a vital component of living organisms.

Water vapour is an effective greenhouse gas whose presence in the atmosphere contributes to keeping climatic conditions comfortable; its absence would lead to a drastic drop in temperature, transforming the Earth into a cold, inhospitable planet. Water is a precious component of the biosphere and plays a key role in many geological processes, such as rock weathering, magma formation, volcanic eruptions, and the accumulation of mineral deposits; its occurrence within the upper mantle is responsible for the soft mechanical properties of the asthenosphere that enable plate tectonics. For all these reasons, the geochemical cycle of water has a paramount geological significance that has implications for the evolution of the planet, the equilibrium of the external environment, and the origin of life.

The global cycle of water—and of all the natural chemical components—consists of two subsidiary cycles (Fig. 7.1). One takes place at the Earth's surface (the **external** or **shallow** or **exogenic cycle**), whereas the other involves the crust and mantle (**deep cycle**). The external cycle takes relatively short times, from days to thousands of years; for this reason, it is also referred to as the **short cycle**. The deep cycle takes tens to several hundreds of millions of years and is also indicated as the **long cycle.**

The exogenic cycle comprises three major steps: the transfer of marine and continental water to the atmosphere through evaporation, returning to the sea and land through precipitation, and flowing to the sea of the water that falls on the mainland. In reality, the routes of the exogenic cycle are much more

Water geochemical cycle

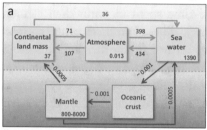
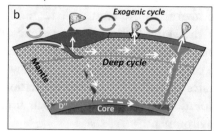

Mass of reservoirs in petatons (10^{15} t) of H_2O; annual fluxes in
teratons (10^{12} t) of H_2O

Fig. 7.1 a Box model of the water cycle. The numbers in the rectangles and on the
arrows are the estimated masses of water in the reservoirs (in petatons) and the
annual fluxes (in teratons) from one reservoir to the other.[1] Colours distinguish the
short and long cycles; **b** Pictorial view of the water cycle. The circular arrows indicate
the exchanges between the atmosphere and the hydrosphere; the white arrows are
the water pathways from the lithosphere to the mantle and back to the surface,
through subduction and volcanism

tangled than just described, also involving plant uptake and transpiration,
sublimation of ice, and the flow of underground water to the sea.

The energy that maintains the water cycle is provided by solar radiation
that causes evaporation; the force of gravity moves water from the elevated
land to the sea. Hydrosphere, atmosphere, and continental landmasses are
the main reservoirs of the exogenic cycle. Many small subsidiary reservoirs
are present within these large reservoirs, including temporary and permanent glaciers, soil, rivers and lakes, underground waters at various depths,
and water in terrestrial and marine living organisms (the **biosphere**).

The deep cycle is less well known and understood, although it plays a
fundamental role in the evolution of the Earth. It develops through the
uptake of water by the ocean floor, its subsequent transportation into the
mantle by subduction, its release to induce partial melting and magma
formation and return to the atmosphere and the oceans through volcanic
activity.

The migration of water from oceans into the oceanic crust occurs both
through the infiltration into rock fractures and the reaction with minerals
and the formation of hydrated phases such as clay minerals, zeolites, chlorite,
or serpentine. The release of water from the subducted crust occurs predominantly in the mantle wedge above subduction zones where it triggers orogenic

---

[1] The prefixes kilo (k), mega (M), giga (G), tera (T), peta (P) and exa (E) express thousands (k =
10^3), millions (M = 10^6), billions (G = 10^9), thousands of billions or trillion (T = 10^{12}), millions
of billions or quadrillions (P = 10^{15}) and billions of billions or quintillions (E = 10^{18}), respectively.

magmatism; a very minor fraction mixes with the mantle and then is bought back to the surface by volcanoes both along diverging plate boundaries and in intra-plate environments.

Therefore, the deep cycle is regulated by subduction and magmatism; these processes operate in opposite directions and continuously recycle water between the inside and outside of the Earth—a journey that takes millions to hundreds of millions of years.

Oceans contain a volume of 1.37×10^9 km^3 and a mass of 1.39×10^{18} metric tons (1.39 exatons), which is about 97% of the entire hydrosphere. Glaciers and permanent snow (2.0%), underground storage (0.68%), atmosphere, rivers, and lakes are much smaller reservoirs (see Table 2.1), but are important for water supply because of their low salinity. Living organisms contain about 1.1×10^{12} metric tons of water, which is equal to 0.000079% of the hydrosphere; a precious but negligible fraction compared to the Earth system as a whole.

The residence time of water in the shallow reservoirs is relatively short; it ranges from an average of approximately ten days for the atmosphere to weeks or months for rivers, years for lakes, a few thousand years for oceans, and tens of thousands of years for ice caps and certain underground 'fossil' waters. Fluxes among various reservoirs are very variable but relatively fast.

The short water cycle is very sensitive to external disturbances. Irrigation and the construction of dams curtail the annual flux from the land to the sea, and the enhanced evaporation it causes has fairly immediate effects on the local climate. The removal of vegetation has the opposite outcome since it reduces transpiration. All these imbalances can be resolved rather quickly once the causes of perturbation are removed, because rapid fluxes, relatively small masses involved, and short residence times rapidly restore the previous equilibrium.

The characteristics of the deep cycle are less known. Water contents in the upper mantle are estimated to be in the range of 0.02 to 0.2% of the total mass (i.e. 200 to 2000 ppm or grams of water per ton of rock). However, much larger amounts (around 2.0–3.0% by weight) are stored in the transition zone by the high-pressure minerals wadsleyite and ringwoodite.[2] Overall, because of its large volume, the entire mantle is thus estimated to contain about 10^{19-20} tons of water, a much greater amount than contained in all the Earth's oceans.

The flux of water from the oceans to the mantle through subduction is estimated at approximately 0.9 to 1.9 gigatons per year. A similar or

---

[2] Huang et al. [1], Hirschmann and Kohlstedt [2].

perhaps slightly lower amount is lost by volcanism.[3] Residence times of water in the mantle are extremely long and variable; the shorter time is typical of the mantle wedge above subduction zones where water is brought back to the surface by orogenic volcanism after a few million years. By contrast, water distributed throughout the rest of the mantle is stored for hundreds of millions of years before being brought back to the surface through magmatism.

The deep cycle is a fundamental process governing the long-term evolution of the planet. The outward flux through volcanism would ultimately lead to the complete loss of water from the mantle and its accumulation on the Earth's surface (Chap. 3). Since the water content of the Earth's mantle is much greater than that of the hydrosphere, the obvious consequence would be the formation of vast oceans, with little to no subaerial land. Subduction prevents this outcome by continuously bringing water back to the mantle and establishing a dynamic balance (**steady-state**) between the superficial and deep reservoirs: this is indeed a paramount result of plate tectonics that has vital consequences on the Earth system.

7.4   The Sodium Cycle

Sodium (Na) makes up about 2.5% by weight of the Earth's crust; large quantities (about 1.5×10^{16} metric tons) occur in the seawater. Because of its high reactivity, sodium does not exist in the free state in nature but is bound to other elements to form various compounds such as the sodium aluminium silicate mineral plagioclase (**albite**: $NaAlSi_3O_8$) and sodium chloride (**halite**: $NaCl$), the everyday table salt. Plagioclases are among the most abundant minerals of igneous rocks (Chap. 1); **halite** or **rock salt** is formed by precipitation from seawater when subjected to intense evaporation in natural evaporite basins (Chap. 2) or at saltworks.

As with water, a short shallow cycle and a long deep cycle are recognized for sodium (Fig. 7.2). The short cycle starts from rock weathering that removes sodium away from minerals and delivers it to rivers and seawater; successively, the element is returned to the continents through marine aerosol and coastal rainfall, or the precipitation of halite in evaporite basins.

The deep cycle begins with the migration of sodium from the seawater into the oceanic crust through water–rock chemical interaction. Subduction of the oceanic lithosphere then transports sodium into the mantle. Finally, magmas

---

[3] Parai and Mukhopadhyay [3].
   Genda [4]

Sodium geochemical cycle

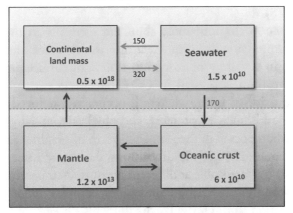

Mass of reservoirs and annual fluxes in milion tons or megatons (10^6 t) of Na

Fig. 7.2 Simplified geochemical cycle for sodium. The numbers in the rectangles indicate the amounts of Na in the reservoirs; the numbers on the arrows are the annual fluxes of sodium between the land and seawater, and from seawater to rocks of the ocean floor

formed by mantle melting extract sodium from peridotite and transfer it back to the Earth's surface, thus completing the global cycle.

The key parameters of the short cycle in Fig. 7.2 show that the amount of sodium passing from the continents to seawater is about 320 million tons per year; the annual return flux is much smaller, amounting to about 150 million tons. This large difference denotes a net excess of sodium being annually transferred to the sea. Such an unbalance should result in a progressively saltier ocean that should transform, after about 100–150 million years, into an immense brine inhospitable for life, or at least for a large part of it (Box 7.2). However, this does not happen because the excess amount of sodium is taken up by the oceanic crust and then transported from the Earth's surface back into the mantle. In other words, the continuous accumulation of sodium in seawater is prevented by plate tectonics, again documenting its paramount role in regulating the Earth system and favouring life.

7.5   The Carbon Biogeochemical Cycle

Knowledge about the biogeochemical cycle of carbon is necessary not only for satisfying the intellectual curiosity about the functioning of the Earth system, but it is also pivotal for a scientifically based understanding of the changes experienced by the Earth's environment during the past and at present (Chap. 8).

Carbon is one of the four most abundant elements in the solar system. The carbon atom contains four electrons in its outer shell and, therefore, can exist in a variety of oxidation states. In surface and near-surface environments, carbon is present in CO_2, bicarbonate ions, and carbonate minerals in an oxidised form, whereas it occurs as a reduced form in many organic compounds such as proteins, carbohydrates, and hydrocarbons. In the mantle, carbon occurs in the elemental form of graphite or diamond and in various compounds (CO_2, hydrocarbons, carbonate minerals and melts, metals carbides) that are stable at different pressures, temperatures, and oxidation (**redox**) conditions. Carbon undergoes complex redox modifications during its migration through the Earth's domains, several times passing from organic to inorganic compounds, and *vice versa*.

The geochemical cycle of carbon affects both the shallow and deep spheres of the Earth. The shallow (short) cycle reservoirs include the atmosphere, hydrosphere, terrestrial and marine organisms, and soil; crustal sediments and the mantle are the main reservoirs of the deep (long) cycle.

The pathway of carbon through the Earth's reservoirs is quite complex, and perhaps this is the primary reason the carbon cycle is overlooked by popular science. Some effort is required to follow it on the simplified diagram of Fig. 7.3, which describes the carbon cycle during the Quaternary period, before the industrial revolution.

From the point of view of humans and the entire biosphere, the atmosphere is the central reservoir. In the terrestrial environment, green plants employ the energy from sunlight to perform photosynthesis, producing carbohydrates and molecular oxygen out of atmospheric carbon dioxide and water.[4] Subsequent reactions convert carbohydrates to a variety of organic compounds that are transferred to animals through feeding. A fraction of carbon returns from animals to the atmosphere as CO_2 through breathing. Another fraction is transferred to the soil by falling leaves, excrement, and the decomposition of dead organisms. A large quantity of such carbon returns from soil to the atmosphere as methane or carbon dioxide through the decay of organic material; a minor fraction is delivered to rivers and the oceans, or is buried in sediments, i.e. in a reservoir of the long cycle.

In the **marine environment,** there is a continuous exchange of carbon dioxide between the atmosphere and hydrosphere. Marine organisms use carbon from seawater while living, but accumulate it on the seafloor after

---

[4] The reaction of the oxygenic photosynthesis can be simplified as: $6CO_2 + 6H_2O +$ sunlight energy $\rightarrow C_6H_{12}O_6 + 6O_2$.

Pre-industrial carbon biogeochemical cycle

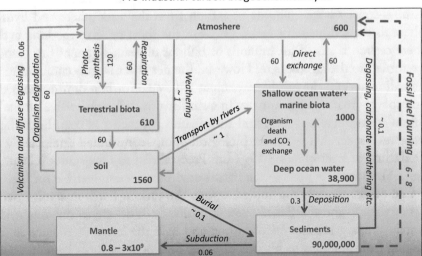

Mass of reservoirs and annual fluxes in gigatons (10^9 t) of carbon

Fig. 7.3 Simplified model of the carbon biogeochemical cycle. The masses and fluxes of the short cycle refer to pre-industrial time, apart from the dashed red line on the right side of the diagram that has been added to show the present-day anthropogenically driven annual flux of carbon from sediments to the atmosphere through the consumption of fossil fuel and cement production

death by depositions of their bodies, finally transferring carbon from oceans to the sedimentary reservoir.

A fraction of the sediments deposited on the ocean floor is involved in subduction and brings its carbon to the mantle[5]. The return of carbon to the atmosphere is provided by CO_2 degassing at active volcanoes or along deep faults.[6]

A large proportion of sediments, especially those of continental shelves or accretionary prisms, are not involved in subduction and can be either buried and heated by metamorphism, liberating CO_2, or uplifted by tectonic movements and exposed at the Earth's surface, releasing carbon directly into the atmosphere through rock weathering.

A crucial step in the global carbon cycle is the uptake of inorganic carbon by marine organisms and its burial on the seafloor as biogenic sediment.

---

[5] Additional carbon is transported to the mantle by the secondary carbonates that form as seawater passes through the basaltic oceanic crust. This step is not considered in Fig. 7.3, for simplicity.

[6] Dasgupta and Hirschmann [5].
   De Paolo [6].

These processes—comprehensively referred to as the **marine biological pump**—convey large amounts of carbon from the atmosphere and hydrosphere to the sedimentary rocks, i.e. from the short cycle to the long cycle, where carbon is stored for millions or billions of years. Some of this carbon is returned to the atmosphere. However, the net effect is an accumulation in the sediments and the nearly permanent removal of carbon dioxide from the atmosphere-hydrosphere domain. In a similar way, terrestrial plants also take carbon from the atmosphere and turn it into sediments as coal. Both these processes, especially the marine biological pump, contributed significantly to the decline of atmospheric CO_2 from Precambrian to the present (Chap. 2, Fig. 2.1), ultimately making the Earth's environment comfortable for life.

Carbon is unevenly distributed among the various Earth reservoirs. Sedimentary rocks and the Earth's mantle, which both belong to the long cycle, contain the largest amounts of carbon and have the longest residence times, ranging from millions to billions of years. Sedimentary rocks contain about 90 million gigatons of carbon, 70 million of which are in carbonates (inorganic carbon) and 20 million in kerogens, hydrocarbons, and coal (organic and elemental carbon). The carbon content of the mantle is estimated to range from 0.8 to 3 billion gigatons. About 4×10^9 gigatons are probably stored in the core. The annual fluxes in and out of the deep reservoirs are very slow.

The short cycle comprises much smaller carbon reservoirs, but the fluxes are notably higher. Residence times in the various reservoirs are very short at the geological time scale. On average, CO_2 molecules remain in the atmosphere for a few years, a few decades in terrestrial vegetation, a maximum of centuries in the upper soil and the shallow ocean water, and a few millennia in the lower soil horizons and the deep sea.

The environmental implications of the carbon cycle will be discussed in Chap. 8. For the time being, it is worth introducing here the implications of the contrasting characteristics of the shallow and deep cycles.

The small size of the short cycle reservoirs makes them prone to modification by external events. However, large fluxes favour circulation, thus quickly re-distributing carbon and levelling out unbalancing. By contrast, the reservoirs of the long cycle are enormous, have very long residence times, and relatively slow fluxes; therefore, they are very stable and require geological times to be modified. These contrasting characteristics make the short and the long cycle two largely autonomous, although not independent systems. Their reciprocal interaction is extremely slow and takes place over geological time. Any alteration of the exchanges between the long and the short cycle, such as a forceful acceleration of carbon flux from the deep reservoirs to the shallow ones (e.g. from sediments to atmosphere-hydrosphere), has dramatic consequences that require geological times to be recovered.

Unfortunately, the carbon flux from sediments to the atmosphere has undergone a substantial increase during the last two centuries through the release of CO_2 from the burning of fossil fuels and the calcination of marl and limestone to produce cement. The anomalous flux, represented as a dashed line in Fig. 7.3, has raised the atmospheric carbon content from about 600 gigatons during pre-industrial times to the present-day value of 850 gigatons. Such a sharp increase is a major concern of both the scientific world and the general public because of the implications for global climatic change, as will be discussed in Chap. 8.

7.6   The Phosphorus Biogeochemical Cycle

Phosphorus (P) is the eleventh most abundant element in the Earth crust, where is mainly contained in apatite—$Ca_5(PO_4)_3(OH,Cl,F)$—an accessory mineral that is sometimes concentrated in certain sedimentary and magmatic rocks. Most of the phosphates employed in agriculture and industry came from these deposits; accumulations of bird or bat excrement (**guano**) are a minor but important local phosphorus resource in some regions. There are no gaseous compounds of phosphorus. Therefore, it is absent in the atmosphere; however, it occurs in the microscopic dust particles transported around the world by winds, thus providing an important source of nutrients for the oceans and remote islands distant from the continents. Phosphorus is an essential element for living organisms, present in DNA and RNA, phospholipids of cell membranes, and the adenosine triphosphate (ATP) that provides energy to cellular reactions and is indispensable for metabolism. It also makes up the hydroxyapatite that occurs in the hard tissue of bone and teeth.

Phosphorus is very scarce in natural waters because of the low solubility of its compounds. Since only dissolved phosphorus can be assimilated by plants (**reactive phosphorus**), low concentrations are a limiting factor for the development of terrestrial and aquatic vegetation. As perceptively stated by Isaac Asimov, phosphorus is a 'bottleneck' for life.

The biogeochemical cycle of phosphorus can be assumed to start from the alteration of apatite in rocks, which brings phosphate into the soil, where it is available for use by vegetation. Phosphorus is then transferred from plants to animals, returning to the soil through falling leaves, excretions, and organic decay. A significant fraction of the element is conveyed as reactive phosphorus from soil to seawater, where is used by phytoplankton and passed to zooplankton and, hence, to the higher levels of the marine food chain. The deposition of the soft components of marine organisms carries phosphorus to seafloor sediments. Later, these sediments are either raised by tectonic movements and exposed at the Earth's surface or involved in subduction and transported into the upper mantle. Here, phosphorus is collected by magmas during partial melting and brought back to the surface.

The low solubility of phosphorus compounds is a severe problem for agricultural production, which has been traditionally addressed through the use of phosphorus-rich substances, such as manure and human excrement, for fertilizer. The shortage of phosphorous might also explain why life took so long to develop in the course of Earth's history. However, the scarcity of phosphorus in natural waters has certain positive aspects as it prevents uncontrolled algae explosions whose aerobic destruction by bacteria depletes the water of oxygen, causing anoxic conditions and fish deaths (**eutrophication**).

7.7 The Nitrogen Biogeochemical Cycle

Nitrogen is one of the five most abundant elements of the solar system, the major component of the Earth's atmosphere (4×10^{15} tons, and about 78% by volume), and an essential constituent of living organisms. Its concentration in the lithosphere is low, at hundreds to a few thousand parts per million.[7]

Atmospheric nitrogen occurs as diatomic molecules (N_2) and, to a much lesser extent, in the harmful compounds N_2O and NO_x, which contribute to acid rain, stratospheric ozone destruction, and the greenhouse effect. In the terrestrial and marine environments nitrogen occurs as N_2, NO_3^-, NH_3, NO_2^-, NH_4^+, and organic compounds. Complex speciation depends on ambient redox conditions in different geological domains.

Although immersed in a sea of nitrogen, vegetation can minimally absorb the element for its growth. This paradox comes from the stability of atmospheric nitrogen molecules, whose two atoms are held together by a strong triple bond. Therefore, it is necessary for nitrogen to be transformed into NH_4^+ and NO_3^- by electrical discharges, fires, or bacterial activity in order to be transformed into a state that permits nitrogen utilization by terrestrial vegetation or marine phytoplankton. Industrial fertilisers are treated to provide nitrogen in a form that makes it readily available for plant uptake.

---

[7] Canfield et al. [7].
  Mysen [8].

The biogeochemical cycle of nitrogen is complex. The atmosphere is the central reservoir. Combustion, electrical discharges in the atmosphere, and organisms in the soil transform nitrogen into various compounds that are utilized by vegetation and then released back to the atmosphere as molecular N_2. The marine cycle is regulated by the arrival of nitrogen from rivers or directly from the atmosphere. A fraction of marine nitrogen returns to the atmosphere as N_2; another fraction is fixed by organisms and subsequently deposited in ocean sediments, thus passing from the shallow to the deep cycle. Some sediments are transported into the mantle through the lithospheric subduction, where the element is incorporated into magmas and then brought back to the surface by magmatism.

The nitrogen cycle has been subjected to severe changes during geological time. Early Earth history was dominated by reduced forms of nitrogen, such as NH_3. In aquatic systems, a major modification occurred during the transition from anoxic to oxic conditions in the Palaeoproterozoic, when the increased oxygen concentration in the atmosphere gradually affected the oceans, thus favouring the formation of oxidised species, such as NO_3^-. An additional modification has occurred in the last century because of the effects of anthropogenic activities—a topic that will be addressed in Chap. 8.

7.8 The Oxygen Cycle

Oxygen is the third most abundant element in the solar system, after hydrogen and helium. On the Earth, it is a major constituent of the hydrosphere, biosphere, and rocks. It is present in the atmosphere as free diatomic molecules, amounting to 6×10^{14} tons and 20.96% of the total atmospheric volume. Its cycle involves all domains of the Earth, and oxygen interacts with the cycle of other elements and substances such as carbon and nitrogen.

Molecular oxygen in the atmosphere is formed through photosynthesis and photochemical processes. Photochemical reactions (**photolysis**) occur in the stratosphere, where water molecules struck by ultraviolet rays split into their constituent atoms—hydrogen and oxygen. Because of its low atomic weight and gaseous state, hydrogen tends to disperse into space and, therefore, does not recombine with the oxygen that remains in the Earth's atmosphere, forming diatomic molecules (O_2). Some oxygen molecules can then react with free atomic oxygen to form ozone (O_3).

The present rate of production of photochemical oxygen is about 10^5 tons per year. Photosynthesis produces about 10^{11} tons per year, a million times more. More than half of the photosynthetic oxygen is produced by marine phytoplankton; the other fraction is generated by terrestrial vegetation. Consumption of oxygen occurs by breathing, degradation of organic matter, and reaction with inorganic reducing substances such as volcanic gases (H_2S, SO_2, CH_4, H_2) and some mineral-forming elements (e.g. Fe^{2+}).

Photosynthetic oxygen production began in early Achaean time. However, its abundance in the atmosphere was negligible because continuously consumed by methane (CH_4), ammonia (NH_4), ferrous iron (Fe^{2+}), and other reducing elements and compounds present within both the fluid and solid Earth. However, a substantial fraction of oxygen participated in the subduction process as a component of water and various minerals in seafloor sediments and oceanic crust. The mantle, therefore, became gradually oxidised over geological time, releasing progressively lower quantities of reducing chemical species through volcanic activity. Such a reduction resulted in lower consumption of atmospheric oxygen and its accumulation in the atmosphere, finally leading to the **Great Oxidation Event** in the early Proterozoic (2400–2100 years ago) when concentrations reached values around 2% by volume—another major consequence of plate tectonics. An alternative view is that oxygen variation is related to the rate of the spin of Earth around its axis. It is well known that the young Earth rotation was much faster than it is today. Slowdowns in Earth's rotation spin followed a stepped pattern, which mirrors quite closely that followed by the increase in atmospheric oxygen. These parallel patterns suggest that additional continuous hours of daylight boosted photosynthetic microbes, enhancing the production of oxygen.[8]

7.9   The Iron Cycle

Iron is the main chemical component of the Earth. It is concentrated in the core, together with other metals (Ni, Co, Au, Ir, Pd, etc.), which, for this reason, were called siderophile (iron-loving) by Victor Goldschmidt. The concentration of iron in the silicate Earth ranges from an average of about 8.0% by weight in the mantle, to around 5.0–6.0% in the continental crust, and 10–11% in the oceanic crust (Chap. 1). High contents in iron make the oceanic crust sufficiently dense to sink into the mantle at subduction zones.

Iron mostly occurs in a divalent 'ferrous' state (Fe^{2+}) in many major rock-forming minerals inside the silicate Earth. On the surface of the Earth and in the hydrosphere, it is present as oxidized ferric iron (Fe^{3+}), which is insoluble

---

[8] Klatt et al. [9].

so that only tiny quantities, on the order of micrograms per litre, are dissolved as Fe^{2+} in water (Chap. 2).

Iron is essential to most living organisms where it is present in the haemoglobin, which transports oxygen from the lungs or gills to the tissues and removes carbon dioxide. It also occurs in many enzymes and is essential for cell growth, immune function, thermoregulation, and DNA synthesis. Iron participates in various physiological and biochemical processes in plants and phytoplankton, especially the synthesis of chlorophyll. Its abundance in the waters is a key factor controlling the growth of phytoplankton.

Simply put, the cycle of iron can be envisaged as starting from the upper mantle, whose partial melting generates Fe-rich basaltic magmas. These magmas rise upward into the crust or are erupted onto the surface. The physical and chemical weathering of crustal rocks generates soil, through which iron passes to vegetation and animals. Much of the transfer from the continents to the ocean is provided by rivers; additional amounts are brought by small solid particles or dust that are taken up by the wind in poorly vegetated or desert regions and transferred to the sea. Submarine hydrothermal activity is also a significant source of iron for oceans.

Iron in the seawater is present as dissolved ferrous iron and as a component of colloids and detrital particles. The latter precipitate rapidly and sink to the seafloor. Iron in other forms remains in solution or suspension, where it is potentially available for use by phytoplankton and other microscopic organisms in their vital cycle. This iron returns to the lithosphere with the accumulation of dead organisms on the seafloor.

The geochemical cycle of iron has changed significantly through geological time. Before the Great Oxidation Event, iron was present mostly in a divalent oxidation state as soluble Fe^{2+} and was dissolved in seawater at high concentrations. The upsurge of atmospheric oxygen caused the progressive oxidation of iron and the formation of insoluble Fe-bearing compounds, which accumulated on the seafloor to form the **Banded Iron Formations** (Chap. 2).

Much of the interest in the biogeochemical cycle of iron stems from its essential role in phytoplankton growth. The occurrence of iron expands the efficiency of the biological pump and stimulates the consumption of CO_2, thus strongly affecting the carbon cycle. It has been suggested that increased iron input to seawater by wind-born dust during the glacial periods expanded the production of phytoplankton and the consumption of atmospheric CO_2, resulting in an anti-greenhouse effect that encouraged, or even caused, cooling (Chap. 8). Intentional introduction of ferrous iron and other nutrients into surface ocean water has been proposed as a mean to stimulate phytoplankton production and, thus, accelerate sequestration from the atmosphere of the excess CO_2 produced by the burning of fossil fuels.

7.10 The Geochemical Cycles of Lead and Arsenic

All the chemical elements participate in the global mass transfer through the Earth's reservoirs. Some of them are essential to life, whereas others are dangerous or lethal (Chap. 8). The list of potentially toxic elements is long and includes aluminium, fluorine, mercury, arsenic, thallium, lead, and many others.[9] As two examples, the geochemical cycles of lead and arsenic are outlined.

Lead has a concentration of a few ppm in the mantle and crust, but it is highly concentrated in massive sulphide ore deposits, along with other metals such as Cu, Cd, Ag, and Zn. Accumulation of sulphides occurs at particular places of the Earth's crust, related to depositions by hydrothermal fluids during late-stage magmatic activity, as described in Box 3.2. Galena (PbS) is the main lead mineral in these deposits.

Lead in the upper mantle is partially transferred to magmas during partial melting and transported toward the crust, where it is released from fluids into rocks around magma chambers. Tectonic uplift and the following supergene processes cause its removal from mineralised rocks and dispersal in soil and water, as well as in the atmosphere through wind-born particles. Lead is then transferred to the biosphere, both in the marine and terrestrial environments, causing harmful effects, as discussed in Chap. 8.

Arsenic is one of the most highly toxic trace elements. It has attracted major attention because of its widespread occurrence in the environment, recently highlighted by improved analytical techniques.

The concentration of arsenic in the mantle is very low, around 0.05 ppm. However, it is strongly concentrated into the liquid during partial melting and thus is transferred to the surface by magmas. As a result, average contents at ppm levels are observed in volcanic rocks. Because of its volatile geochemical character, arsenic concentrates in the magmatic fluid phase and is particularly abundant in the geothermal areas. Arsenic is adsorbed by many minerals and can reach concentration values of hundreds of ppm in some sediments and metamorphic rocks; even higher abundance is found in coal.

---

[9] Sparks [10].

Arsenic is a major constituent of some minerals (e.g. arsenopyrite, realgar, and orpiment), and occurs as minor or trace elements in many others, such as pyrite and apatite that are widely distributed in many rocks. Most of these minerals are poorly soluble under anoxic conditions and, therefore, are virtually non-toxic. However, in the presence of oxygen, they oxidise to release soluble arsenic. As a result, waters can be highly enriched in arsenic. Mining, smelting, coal burning, and industrial activity also dramatically increase concentrations in the environment.

7.11 Summary

The Earth is a large chemical-physical system through which chemical elements move continuously, cyclically migrating throughout the atmosphere, hydrosphere, biosphere, lithosphere, and mantle. These processes—referred to as geochemical cycles—continuously transfer chemical elements, constructing a sort of circulatory system that maintains steady-state conditions for both the internal and external spheres of the Earth.

Each element has its own geochemical cycle; the different cycles continuously interact and interfere, forming a complex mechanism that keeps the Earth system stable. In summary, the cycling of chemical elements can be imagined as a great clock mechanism, in which the wheels of various sizes and materials move at their own speed, but are all interconnected and have interdependent movements that allow the machine to work.

The paths of elements in the superficial environment consist of uninterrupted migrations through external spheres of the Earth—the atmosphere, hydrosphere, pedosphere, and biosphere. The deep geochemical cycles involve the crust and the mantle. The shallow and deep cycles make up two rather distinct geochemical systems that are connected and continuously exchange matter. The circulation of the elements in the shallow cycles is rapid on the geological time scale, whereas the migration through the crust and mantle (deep cycle) is much slower, lasting millions or billions of years. Consequently, the reservoirs of the deep cycles are domains where elements are stored and slowly released in the exogenic environment.

All elements undergo cyclical migration through the Earth spheres. Water evaporates from oceans, condensates, and returns to the hydrosphere after a relatively short time span. Part of the ocean water is involved in subduction, migrates into the upper mantle where it rests for millions of years before being involved in the magmatism, and ultimately returns to the Earth's surface through volcanism. Other elements, such as carbon, have more complex behaviour, affecting both geological and biological systems. For instance, the carbon present in the atmosphere as CO_2 is taken up by green plants—including green algae and land plants—that use it for photosynthesis. Accumulation of dead organisms transfers carbon to sediments, which, in turn, transport it into the mantle through subduction, where it can later become involved in magmatism and brought back to the Earth surface.

Overall, geochemical cycles create an equilibrium for the Earth that regulates its behaviour as a system and establishes the conditions of the external environment. Geochemical cycles have undergone important changes at various times throughout Earth history. The emergence and development of life was a decisive step in the evolution of geochemical cycles. Biological activity significantly modified the composition of the external environment, especially the atmosphere, thereby reshaping the geochemical cycles of several elements, including carbon, oxygen, iron, and nitrogen, among others.

7.12   Box 7.1 The Rock Cycle

The rocks occurring on Earth belong to the three large families—igneous, sedimentary, and metamorphic (Box 1.1). These groups all participate in the Earth's dynamics, transforming one into the other during geological processes. These transformations—referred to as the **lithogenic** or **rock cycle**—are schematically indicated in the box diagram of Fig. 7.4. Since rocks consist of chemical elements, the rock cycle actually results from the combination of their constituent element cycles.

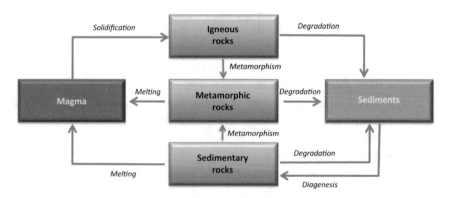

Fig. 7.4 Box model of the rock cycle

Magmatism can be considered as the starting point for the rock cycle. Magmas solidify either inside or at the surface of the Earth to generate intrusive and effusive igneous rocks (Chap. 3). Such a process mostly occurs at the divergent plate boundaries, along the mid-ocean ridges (extensional magmatism); active plate boundaries represent the second most important geological setting for igneous activity (orogenic magmatism). Minor magmatism takes place at specific points inside the plates (intra-plate magmatism).

In the oceanic environment, the igneous rocks basalt and gabbro formed along the oceanic ridges move away from their source zones and go into the mantle at convergent margins, after travelling long distances and having undergone metamorphism on the seafloor and along subduction zones. When temperatures inside the mantle reach the melting point of the metamorphosed rocks, partial melting occurs generating magmas. These rise to the surface to form new igneous rocks, thus closing the loop.

Subaerial igneous rocks undergo weathering and disaggregation (degradation) to produce sediments, from which diagenesis generates sedimentary rocks. These, in turn, may be buried deep inside the Earth, where they are affected by high pressure and temperature, transforming into metamorphic rocks or also undergoing partial melting. Such a transition can happen along the subduction zones, and in the terrains underlying the orogenic volcanoes along the overriding margins.

Sedimentary rocks that remain at the surface undergo a new cycle of weathering and erosion to generate sediments and then sedimentary rocks through diagenesis.

Metamorphic rocks formed along subduction zones or at the margins of the overriding plates can be lifted by vertical tectonic movements and exposed at the surface. Here they are weathered and enter the sedimentary cycle.

In conclusion, the rock cycle comprises a series of transformations that continuously affect rocks, both at the surface and in the interior of the Earth. The physical and chemical weathering on the surface of the Earth destroy minerals and rocks, to produce sediments and ultimately sedimentary rocks. The high temperatures and pressures within the Earth's interior generate metamorphic rocks and magmas, starting from any rock type.

The rock cycle has continuously created, modified, destroyed and ultimately recycled the rocks, during the Earth history. As a consequence, ancient rocks hardly survived, and those that did rarely preserve their original characteristics. Since the rock record is a repository of our planet's history, the effect of the rock cycle severely hampers our capability to reconstruct the past history of our planet.

7.13   Box 7.2 The Geochemical Cycles and the Age of the Earth

The age of the Earth has been a primary concern of scientists and theologists of all civilisations for millennia. The Greek philosopher Aristotle believed the Earth was eternal, whereas epicureans and atomists propended for a non-eternal and young Earth.

Christian scholars accepted the version of creation in the book of Genesis until the seventeenth century and based their estimates of the age of the Earth on what the Bible was reporting.

In 1645, the Irish Bishop James Ussher contributed to the long-running theological debate on the age of the Earth by scrupulously scrutinising the chronology of biblical genealogies and events, finally concluding that the Earth was created in the early morning of October 26th 4004 years BC. This value was not far from other estimates, including those of Johan Kepler and Isaac Newton, two giants we associate with the birth of modern science. However, they shared with Usher the faith in the Bible as a source of scientific information and therefore got similar results in estimating the age of the Earth.

The favour met by the Usher timescale was so broad as to be included in the marginal columns of the authorised version of the Bible translation in English and remained there until the beginning of the nineteenth century.

Many scholars, however, had profound doubts on the age of the Earth estimated from the biblical account simply because time was exceedingly short and could not explain the deeply eroded terrains observed in nature. In the late 1700s, James Hutton (1726–1796), often described as the 'Father of Geology', argued that the Earth's history could be inferred from evidence in present-day rocks. Based on this assumption, Hutton concluded that the deep erosion of Scottish terrains required a much longer time than deduced from the Bible, considering that the Hadrian's Wall of Roman time, which was one thousand years old, had not been that much affected by degradation.

The debate about the Age of the Earth intensified during the 119th century when the new paradigm of **uniformitarianism** was proposed by James Hutton and successively by Charles Lyell (1797–1875). The uniformitarian principle postulates that the same natural laws and processes that operate in the present-day world have operated in the same way during all of Earth history. This concept views geological phenomena as having occurred at the same rate in the past as in the present. Therefore, it has been slowly continuous geological processes, rather than catastrophic events, that have shaped the Earth. Based on this idea, Charles Darwin (1809–1882) calculated that about 300 million years were necessary to erode the rocks originally overlying the Weald, a sandstone formation in Southern England.

On July 1st 1858 a joint paper by Charles Darwin and Alfred Wallace (1823–1913) entitled "*On the Tendency of Species to form Varieties; and on the Perpetuation of Varieties and Species by Natural Means of Selection*" was read at a meeting of the Linnean Society of London, and immediately afterwards published in the "Journal of the Proceedings of the Linnean Society". The following year, in November 1859, Charles Darwin's famous book, "*The Evolution of Species*", was published. The paper and the book presented the new revolutionary theory of evolution by natural selection, which considered the living species occurring on Earth as derived from one or a few ancestors after a numberless series of genetic modifications. Natural selection allowed only the best-adapted species to survive, multiply, and reproduce, whereas other less suitable forms went extinct. Darwin and Wallace avoided mentioning human beings, but, reading their texts made obvious to many that humans were only one of the many species that triumphed in the battle for survival.

It was well known to scholars by the eighteenth-nineteenth centuries, that changes affecting the animal species and eventually transforming them into other species, are exceedingly slow and take millennia, if not longer, to complete. Therefore, the implication for the Darwin-Wallace theory was that several hundred million or even billions of years of evolution were necessary, in order to allow the transformation of a few ancient ancestors into the many current animal species. Such an exceedingly long time required that the Earth be much older than indicated by any available estimate. The opponents of Darwin-Wallace concentrated on this aspect to attack the evolution theory.

In 1868, the influential Scottish physicist William Thomson, later to become Lord Kelvin, calculated that the time needed for the Earth to cool down from its initial state of a molten sphere to the present-day temperature was about 100 million years. Later on, Thomson revised his previous estimate downward to about 40 and then to 20 millions of years.

Darwin and the other followers of the evolution by natural selection were not happy with Lord Kelvin's estimate, arguing that these ages were far shorter than required by the evolutionary theory. We now know that the results of Kelvin calculations were wrong because the phenomenon of radioactivity was unknown. Therefore, Lord Kelvin's estimates could not consider the heat contributed by radioactive decay, an oversight that led to seriously underestimating the planet's age.

Unfortunately, calculations grounded on geological criteria reached similar results as Lord Kelvin. In 1868, the director of the Scottish Geological Survey Archibald Geikie, concluded that Earth was not older than 100 million years, based on the amount of erosion and its effects on the landscape. Another method employed the rate of sedimentation and the total thickness of sedimentary sequences accumulated in the oceans. Based on the uniformitarian principle, it was calculated that the accumulation of sediments at the current rate would have brought to the total filling of the ocean basins in a few tens of million years.

A rather rigorous method to estimate the age of the Earth was based on the mean salinity of the oceans, an approach early advocated by the polymath Edmond Halley in the eighteenth century. In the famous work entitled "*An estimate of the geological age of the Earth*" published in 1899 in The Scientific Transactions of the Royal Dublin Society, the chemist and geologist John Joly of the Dublin University considered the concentration of sodium in the seawater and the rate rivers transport the element to conclude that the current level of marine salinity would be reached after 99 million years, which was then believed to be the age of the Earth. Joly assumed early ocean water was virtually free from salts, that sodium had accumulated continuously since the formation of the planet, and there had been no significant loss from seawater over time. Obviously, Joly could not know that sodium was partially absorbed by seafloor rocks and recycled through the Earth's mantle as described in this chapter.

The matter of the age of the Earth was finally settled at the beginning of the twentieth century after the discovery of radioactivity in 1896 by Henri Becquerel. Radioactive decay allowed the absolute age of minerals and rocks to be determined from which a geological time scale has been developed and continuously refined (Box 9.1). In 1905, Ernest Rutherford used the radioactive decay of uranium isotopes to infer that the Earth was at least 500 million years old. Successively, studies by Bertram Boltwood, Robert Strutt, and Arthur Holmes of uranium isotope decay in lead minerals yielded ages extending to billions of years. The latest estimate of 4.567 million years is that of Claude Allégre and co-workers, who measured this age in some inclusions of the Allende meteorite (Chap. 9). Such a date is often assumed as the birth date of the Earth.[10]

Although the magnitude of Earth's lifetime has become clear since the beginning of the last century, the reasons why the oceans were not filled with

---

[10] Meteorites rather than other geological samples are used to determine the timing of the accretion of the Earth. Rocks and other material cannot provide a direct date because they have been destroyed or extensively modified by the dynamic geological processes associated with the Earth's dynamics.

sediments and the seawater was not a brine remained a geological puzzle for decades. The arrival of the plate tectonic theory in the 1960s finally shed light on this issue by making it clear that the sodium cycle is not restricted to just the oceans and the mainland, as assumed by Joly, but involves deep reservoirs that take up the element from the seawater and sequester it within the Earth for millions or billions of years. As for sediments, their scarcity in the ocean was the obvious consequence of continuous seafloor spreading, seafloor subduction, and accretion along converging margins.

To conclude, the quantitative calculations by scientists of the nineteenth century are commendable because they represented a new and revolutionary approach to quantify the timing of global geological processes. In hindsight, they were erroneous because of incorrect assumption about the factors controlling the physics and chemistry of the Earth and a very rudimental knowledge of geochemical cycles. With regard to the Joly's estimate, the date of 99 million years was not the age of the Earth but rather the time for sodium to reach a steady-state condition or the amount of time needed to replace all sodium in the ocean, i.e. its residence time in the seawater. The current estimate of this parameter is about 70 million years, not that different from Joly's calculation.

References

1. Huang XG, Xu YS, Karato SI (2005) Water content in the transition zone from electrical conductivity of wadsleyite and ringwoodite. Nature 434:746–749
2. Hirschmann MM, Kohlstedt D (2012) Water in the Earth's mantle. Phys Today 65:40–45
3. Parai R, Mukhopadhyay S (2012) How large is the subducted water flux? New constraints on mantle regassing rates. Earth Planet Sci Lett 317–318:396–406
4. Genda H (2016) Origin of Earth's oceans: an assessment of the total amount, history and supply of water. Geochem J 50:27–42
5. Dasgupta R, Hirschmann MM (2010) The deep carbon cycle and melting in Earth's interior. Earth Planet Sci Lett 298:1–13
6. De Paolo (2015) Sustainable carbon emissions: the geologic perspective. Mater Res Soc Energy Sustainab: A Review Journal 2:1–16
7. Canfield DE, Glazer AN, Falkowski PG (2010) The evolution and future of Earth's nitrogen cycle. Science 330:192–196
8. Mysen B (2019) Nitrogen in the Earth: abundance and transport. Progr Earth Planet Sci 6:1–15
9. Klatt JM, Chennu A, Arbic BK, Biddanda BA, Dick GJ (2021) Possible link between Earth's rotation rate and oxygenation. Nat Geosci 14:564–570

8

The Geochemical Cycles and the Environment—How Man is Changing the Earth

We forget that the water cycle, the air cycle and the life cycle are one.
Jacques Cousteau (1910–1997)

8.1 Introduction

Geochemical cycling of the elements has been going on since the beginning of Earth history. The way these cycles have operated has evolved slowly with time, although dramatic upsetting has occurred from time to time because of catastrophic events. The planet has always recovered from disasters, demonstrating that the Earth system can absorb anomalies, albeit very gradually, over geologic time scales.

Over the last thousand years known as the **Anthropocene**[1] human activities have significantly contributed to the modification of geochemical cycles. Man's impact on the environment began in prehistoric times and has progressed in an ever faster exponential rate during the last decades. Modification of geochemical cycles has resulted in environmental disturbances that are harmful to humans and animals worldwide.

---

[1] Anthropocene is the proposed geological epoch beginning with significant impact by humankind on the Earth and its ecosystems (Chap. 9).

Many elements and compounds, including phosphorus, nitrogen, fluorine, sulphur, lead, mercury, aluminium, uranium, have undergone severe alterations of their geochemical cycles. Anomalous quantities of these elements are sometimes naturally released from rocks and minerals in highly enriched mineralised regions. However, most of the time, environmental pollution is intentionally or inadvertently related to human activities.

The incorrect or excessive dispersion of harmful substances into the environment is a far-reaching and complex issue. However, it basically consists of the perturbation of natural cycles of the elements and is, in its very essence, a geological problem. Therefore, environmental problems are a crux subject for study by geological sciences whose understanding requires the knowledge of how the elements behave in nature and the way they are affected by human activities.[2]

This chapter briefly addresses this broad topic, focusing only on a few example elements. A particular emphasis will be given to the modification of the geochemical cycle of carbon and its effects on the atmosphere and ocean—one of the most widely debated issue at present among specialists and on the media.

8.2   Geochemical Cycles and Environmental Pollution

The enormously growing world population and the increased production of technological devices require massive amounts of food and metals. The additional needs have expanded the extraction and consumption of natural resources, conveying into the hydrosphere, atmosphere and soil enormous quantities of metals and other substances whose accumulation is resulting in severe environmental pollution.

8.2.1   Toxic Elements

Pollution by arsenic is a primary health problem. Arsenic is a carcinogenic element whose toxicity in the waters is significant even at concentrations of a few micrograms per litre. Arsenic has been widely used as a component of insecticides, herbicides and wood preservatives until quite recently. Historically, yellow orpiment and orange-red realgar were used as pigments and are commonly found in many paintings, ceramics, and even wallpapers. Its use

---

[2] Harmon and Parker [1].

has greatly declined in the last decades, but it is still needed in the modern electronic industry to construct semiconductors.

Water contamination by arsenic can be caused by both anthropogenic and natural processes. Pollution related to human activity is more common in industrialized countries, typically deriving from chemical industrial waste and the use of pesticide. Most times, arsenic is a natural pollutant directly provided by rocks, especially those hosting underground aquifers. As explained in Chap. 7, the solubility of arsenic is tightly related to its oxidation state. Therefore, the release of arsenic from aquifers is a consequence of deep drilling and lowering of water table that speed up the oxidation process and favour the liberation of the element into solution. Natural arsenic pollution occurs in many countries. Problems are severe in places such as West Bengal and Bangladesh, where concentration of arsenic in the water extracted from shallow wells is one order of magnitude higher that the limit of 10 μg per litre recommended by the World Health Organization. However, anomalous arsenic concentrations in drinking water have been found at several places in the United States and Europe, and sometimes are contained in mineral waters that are bottled and sold in the market.

A similar story holds for other toxic elements, such as lead and mercury. Lead has no biological function and is potentially dangerous for humans and other living organisms. Its toxicity involves a harmful impact on several organs, the nervous system, and blood pressure. However, the scarcity in the rocks does not raise any significant ecological problem, except for mineralised areas. Unfortunately, the metal has been exploited by humans for millennia and used for a wide range of applications. Ancient Egyptian cosmetics were prepared by powdering lead minerals such as galena and cerussite, and the Romans used lead for water piping; both practices survived until quite recently.

Although currently prohibited for many industrial uses, lead metal has been extensively used in paint, pewter, glass and crystal, solder, as an anti-knock agent in gasoline, and is still in use today for batteries, small arms ammunition, glass, ceramic glaze, and radiation protection.

Mercury is another highly toxic element that affects the nervous system and the human brain, especially for children. It is present in many rocks in considerable amounts and also a constituent of volcanic gases, which readily introduce the element into the environment. Most of the pollution problems today are associated with small-scale gold mining, which makes use of mercury to extract gold, exploiting the capability of the element to bind to precious metals. Other sources of pollution include the combustion of coal for electricity generation and limestone for cement production.

Mercury can enter the oceans through river water or other sources. Once in the water, it undergoes a number of reaction and enters the marine food web. Thus, consumption of fish can bring mercury to terrestrial animals, including humans.

Overall, the environmental problems caused by toxic metals are, geologically speaking, related to their geochemical cycles. Most elements are naturally accumulated in particular places of the Earth's crust by sedimentary, metamorphic or magmatic processes (see Box 3.3). Subsequent tectonic movements lift the mineralised rocks to the surface, where elements slowly enter the exogenic cycles. These are natural sources, which have provided chemical components to the external environment for millions of years, possibly producing some effects only at local scales. However, the intensive exploitation of mineral deposits by humans has enormously accelerated the element transfer from the solid Earth to the external environment, creating disequilibria that will take geological times to recover.

8.2.2 Phosphorus and Nitrogen Compounds

Food is a primary necessity that needs enormous amounts of fertilisers. Phosphates and nitrogen compounds are essential to boost agriculture.

Phosphates are extracted from sedimentary deposits or, to a minor extent, magmatic reservoirs, both belonging to the long geochemical cycle of the element. Therefore, the use of phosphates in agriculture results in a forced and rapid transfer of phosphorous from the long to the short cycle. Extensive use of fertilisers has more than doubled the flux of phosphates to the sea, compared with the natural cycle. Discharges by humans and animals have added several million tons per year to natural waters. Available data show that only half of the approximately 15 million tonnes of phosphorus used annually in agriculture is absorbed by plants for their growth, while the other half is washed away by water and added to rivers. It should also be considered that about 30–40% of the food produced annually undergoes destruction or degradation, which leads to the loss of another million tonnes of phosphorus per year.

The solution to these anomalies can only be provided by more rational use of resources, which is more respectful of the natural biogeochemical cycle of phosphorous. More efficient practices in agriculture, a smarter distribution of food, and the recovery of phosphorus linked to human and animal excretions could limit anthropogenic anomalies. All these actions affect the short cycle and are potentially capable of strongly reducing the extraction of the element from the non-renewable reserves of the long cycle. It has been calculated that recycling all manure and human excreta could halve the global extraction of phosphate from the lithosphere.

Excess of **nitrogen** compounds in the environment represents a major pollution issue that is not adequately considered by the regulatory authorities.

Nitrogen compounds are essential for several purposes, including electronics, food packaging and preparation of rocket propeller and ammunition. However, major use (urea, nitrates, sulphates of ammonia, etc.) is in agriculture as fertilisers.

The addition of synthetic nitrogen to soils is a major source of pollution. Some 15 million tons of nitrogen are added to crops, but only 25–30% of this amount is utilized; therefore, the excess ends up in groundwater or volatilised into the atmosphere. However, nitrogen fertilisers are not extracted from the lithosphere as with phosphorous, but are synthesised from elements mostly coming from the atmosphere through the Haber–Bosch process, an industrial procedure named after the German chemists that developed it at the beginning of the 20th century. Therefore, perturbation of the nitrogen geochemical cycle by agricultural practises consists of the forceful transfer from the atmosphere to pedosphere and hydrosphere. Excess use of nitrogen compounds results in anomalous accumulations in soil that then transfers to the hydrosphere to affect water quality and cause eutrophication. Another source of nitrogen pollution results from the burning of fossil fuel and calcination of limestone for cement production that together generate some 25×10^6 tons of nitrogen oxides (NO_x) per year.[3] Geologically speaking, this is a forceful transfer of nitrogen from the lithosphere to the atmosphere.

Human activities have more than doubled the input of nitrogen compounds to the environment, interacting heavily with atmosphere, hydrosphere, pedosphere, and biosphere. The complex nature of nitrogen pollution, its multi-source origin, and the many sectors that would need to be involved to affect a solution make nitrogen pollution a difficult issue. A limitation on fossil fuel consumption, wastewater treatment and improved agricultural practices—e.g. matching the use of fertilisers to crop requirements and better assessing the dose of fertilisers and modifying the dosage as a function of weather—are necessary to address the problem.

---

[3] Nitrous oxide (N_2O), better known as laughing gas, that is used in hospitals and dental clinics as an anaesthetic, is one of the most dangerous nitrogen gaseous compounds. Having a 300 times stronger greenhouse power than CO_2, it contributes significantly to global warming. It also contributes to ozone consumption and acid rain. Its concentration in the atmosphere has increased from about 260 parts per billions (ppb) during the pre-industrial time to the present-day concentration of about 330 ppb.

8.3   Environmental Aspects of the Carbon Cycle

One of the most impactful global environmental issue arises from the disruption of the geochemical cycle of carbon.[4] The widespread disposal of plastic wastes into the environment, plus the introduction into the atmosphere of excessively large amounts of carbon dioxide (CO_2) produced by the burning of fossil fuels, are two anthropogenic activities currently distorting the natural carbon cycle. All these anomalies basically result from a rapid and massive relocation of carbon from the huge reservoir of the long cycle (the sedimentary rocks) to a small reservoir of the short cycle (the atmosphere). Because of the impact on ocean chemistry and global climate, the accumulation of CO_2 in the atmosphere is attracting much of the attention from both the scientific community and the public media. But the dispersion of plastics and microplastics into the environment is perhaps as much severe problem as carbon dioxide.

Carbon migration from sediments to the atmosphere is one step in the carbon cycle. However, natural fluxes are low, around 0.1 gigatons per year, and this mass of carbon is readily recycled and returned to the long cycle through the biological pump. Human activities have disrupted this delicate dynamic equilibrium by abnormally speeding up carbon flow from the long to the short cycle, with a flux estimated to be around 6–8 gigatons of carbon per year (thick dashed line in Chap. 7, Fig. 7.3). This forceful introduction has transformed the atmosphere composition, increasing CO_2 concentration from pre-industrial values of 260–270 ppm to over 400 ppm. This elevated concentration has resulted in an intensification of the greenhouse effect, and significant modifications in the hydrosphere. Natural processes will take about a hundred thousand years to remove this excess CO_2.

8.3.1 CO_2 and the Ocean Acidification

Any variation of CO_2 in the atmosphere has a consequent effect on the oceans because of the continuous exchange of this gas between the two reservoirs. It is estimated that out of 1300 gigatons of CO_2 added to the atmosphere in the last 200 years by anthropogenic activity, approximately 500 gigatons have been absorbed by the oceans. The consequence of this has been the acidification of the ocean, because dissolved CO_2 in the water forms carbonic acid. Thus, *pH*, the common measure of acidity, has decreased in

---

[4] Suarez et al. [2]

surface seawater by about 0.1 units in the last decades, with a further drop of 0.3 units expected by the end of this century.

Such a decrease in *pH* value appears small, but in fact has already had a profound adverse effect on organisms such as some molluscs, corals and the zooplankton pteropods that have carbonate exoskeletons. This is because this seemingly small acidification dissolves carbonate, thus making organisms' shells more difficult to grow. Corals and pteropods, which have aragonite exoskeletons, are at high risk. Molluscs are among the most diverse and abundant organisms in the ocean; corals form the great reefs that support a quarter of all marine life; pteropods form the base of the marine food web. The decline of these species threatens the entire marine ecosystem.

8.3.2 CO_2 and the Greenhouse Effect

Water vapour is the main greenhouse gas in the Earth's atmosphere, followed by CO_2 and methane (CH_4). These gases let the high-energy radiation from the Sun reach the surface, whereas they absorb the infrared radiation emitted back to space from the Earth. Such a selective filtering favours the accumulation of thermal energy in the troposphere, leading to a hotter global climate—the so-called **greenhouse effect.**

Moderate concentrations of greenhouse gases have positive effects on Earth because they slow down the dissipation of heat into space and preserve the planet's surface from severe cooling. However, their excess prompts abnormal heating of the troposphere. An extreme case is the planet Venus, where the high contents of atmospheric CO_2 (96.5%) contribute to extremely hot surface temperatures of about 460 °C.

Methane is a much more potent greenhouse gas than CO_2. However, it cannot concentrate excessively in the atmosphere because it reacts with oxygen to produce CO_2 and H_2O. Water vapour does not exceed a certain threshold because it condensates as rain or snow when reaching supersaturation. By contrast, CO_2 can rise almost indefinitely, since only the slow reaction with silicate minerals permanently removes it from the atmosphere. Photosynthetic plants and seawater take up CO_2, but ultimately return it back to the atmosphere.

Paleoenvironmental studies show that the Earth's atmosphere has experienced substantial variations in CO_2 content during its history, and these have often been associated with climatic changes. However, the general trend has been negative due to the long-term net transfer of carbon from the atmosphere-hydrosphere to sediments (see Chap. 2).

Starting from the industrial revolution, the trend has been reversed, with CO_2 concentrations rapidly rising, as documented by measurements at many observatories worldwide (Fig. 8.1). A steeper trend has been observed for methane, which has increased from a pre-industrial concentration of about 600 parts per billion (ppb) to the present value of about 1900 ppb. The build-up of greenhouse gases has been accompanied by global warming of about 0.1–0.2 °C per decade, with a general retreat of the Alpine glaciers and sea-level rise by several centimetres. The problem with this issue is not that much the increase in CO_2 and temperature *per se*, but rather the rate at which it is occurring.

There is little room for doubt that the growth of greenhouse gases in our atmosphere is due to human activity, especially the burning of fossil fuels, deforestation, and cement manufacturing. Also, there is little dispute that the rise in temperature parallels the increase in concentration of greenhouse gases. Whether the two processes are causally related is still somewhat contentious.

Most scientists consider that hotter climatic conditions are caused by the increased concentration of CO_2 in the atmosphere and that this trend will continue into the future if the present situation goes on. Others question this causal relationship, suggesting that the present-day climatic variation might

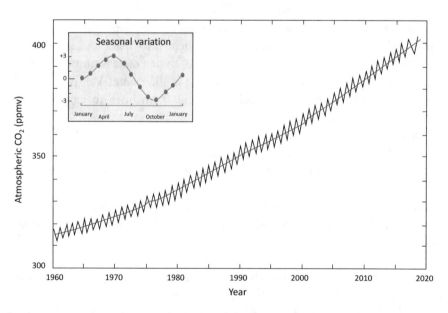

Fig. 8.1 Atmospheric CO_2 variation (in parts per million by volume) over the last 60 years measured at the Mauna Loa Observatory, Hawaii Islands. The inset figure shows the annual oscillations of CO_2 related to seasonal absorption by vegetation in the Northern Hemisphere

be the consequence of factors external to the human activities. There are some eminent geochemists among sceptics.[5]

The anthropogenic impact on climate is one of the thorniest topics of modern Earth Sciences. Therefore, a scrutiny of the history of our planet might provide useful empirical indications to interpret the present and furnish suggestions for the future.[6]

8.3.3 CO_2 and Climatic Change: A Geological Perspective

Much of the history of the lithosphere is contained in rocks and minerals. By contrast, direct information on the atmosphere of the geological past is lacking and its composition can be indirectly inferred from analyses of fossils, rocks, and minerals and geochemical models.[7] However, these proxy studies often yield ambiguous results because many processes may contribute to the present composition of geological materials, making it hard to recognise the effects directly related to paleoenvironments.

8.3.3.1 Precambrian

Geochemical modelling and the few available data suggest that the Early Hadean atmosphere contained moderate amounts of carbon dioxide; successively, concentrations increased through intense volcanism and diffuse mantle outgassing, reaching tens of thousands of parts per millions during the Archaean. However, these high atmospheric CO_2 concentrations did not lead to excessive heating on the Earth surface because of the low energy output from the Sun during the Precambrian (Chap. 9).

---

[5] Allègre [3].

[6] Lear et al. [4].

[7] Indirect information on the past atmospheric CO_2 concentrations is inferred from a number of proxies, including stomata pores on fossil leaves, the boron isotopes of shells of deep-dwelling foraminifera, and the carbon isotope compositions of carbonates in paleosols, microfossils, and algal biomarkers (alkenones). Temperatures are mainly calculated from the distribution of various types of fossils (animals, plants pollens, etc.) inside sediments, and from the oxygen isotope signatures of biogenic materials such as the tests of foraminifera. For the last few hundred thousand years, ice cores drilled in Antarctica and Greenland provide more direct (and reliable) climatic information: CO_2 contents are measured directly from the air bubbles trapped in the ice, while temperatures are calculated from the isotopic composition of water (Box 8.1).

The Proterozoic aeon experienced at least three severe cooling events that enveloped the Earth's surface in an almost continuous ice sheet, from poles to the equator—the so-called **Snowball Earth**. One episode occurred about 2.4 billion years ago; two other global glaciations took place between 720 and 635 million years before present, during the Cryogenian period. Atmospheric compositions during snowball Earth events are poorly known. Some studies suggest that Cryogenian CO_2 concentrations were in the range of a few thousand parts per million and increased significantly at the end of glaciations. Therefore, CO_2 deficiency does not seem the cause of glaciation.

A plausible sequence of steps during the Cryogenian snowball Earth event might have started with a large meteorite impact or a gigantic volcanic eruption that released vast amounts of fine dust particles and sulphate aerosol into the atmosphere, shielding the sunlight, and triggering global cooling and the formation of large ice polar caps. In most instances, the Earth system can recover from such anomalies over a relatively short time. In this case, however, a number of positive climatic feedbacks resulted in a domino reaction (**runaway effect**) that allowed the situation to evolve towards a catastrophic almost complete freezing of the Earth's surface.

In essence, the early formation of polar ice caps amplified the diffuse reflection of solar radiation (**albedo**) encouraging a further drop in the global temperature and thus further inducing an expansion of the ice sheet over the sea. According to the model developed by the Russian climatologist Mikhail Budyko (1929–2001), if the ice cover extends to latitudes of 30°, the albedo becomes so intense as to produce a runaway feedback effect that inevitably leads to the almost complete freezing of the planet. In the case of the Cryogenian glaciations, the climatic runaway was favoured by the paleogeographic position of continents, which were grouped together at low latitudes in the supercontinent **Rodinia**. The equatorial position of the continents left the rocks exposed to the atmosphere while the ice sheet was expanding over the ocean. Consequently, continuing reaction with silicate minerals provided an uninterrupted consumption of atmospheric CO_2, thus favouring further temperature drop and ice expansion. The trend halted and reversed when the ice layer reached the continent and isolated the rocks from the atmosphere, hindering further reaction between CO_2 and silicate minerals. Return to mild conditions was helped by the introduction into the atmosphere of large quantities of volcanic CO_2.

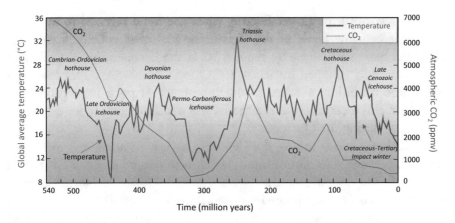

Fig. 8.2 Estimates of temperature and atmospheric CO_2 variations during the Phanerozoic[8]

8.3.3.2 Cambrian to Cretaceous (541 to 66 million years)

Concentrations of atmospheric CO_2 during the Palaeozoic (542–253 Ma) were initially high, perhaps around 10,000 ppm by volume (Fig. 8.2). Multi-proxy estimates suggest that the average global temperature was around 22–24 °C or higher. Successively, developing land plants and the expansion of large forests on the continents led to a dramatic decline in the atmospheric CO_2 that dropped to concentrations as low as 250–300 ppm during the Carboniferous and Permian periods. A long-lived glaciation followed this decline. Another Palaeozoic glaciation is documented in the Late Ordovician, but there isn't any significant CO_2 anomaly during this event.

The Mesozoic (250–65.5 million years ago) was a warm era with little or no ice accumulation. Concentrations of atmospheric CO_2 oscillated between about 2000–4000 ppm and decreased during the Cretaceous period.

8.3.3.3 Paleocene to Miocene (66 to 5.33 million years)

The Earth continued in the warm mode during the transition from the Mesozoic to Cenozoic about 66 million years ago (Fig. 8.3), except for a short-lived temperature drop triggered by the meteorite impact in the Chicxulub area (Mexico) which caused dramatic life changes with the mass extinction at the Cretaceous-Tertiary (KT) boundary. CO_2 concentrations are estimated between 500 and 1000 ppm during the Paleocene. These values increased

---

[8] Scotese et al. [5].

progressively until the Lower Eocene, when the trend reversed to reach the unprecedented low levels of the Pleistocene-Holocene epochs. Simultaneously, the temperature rose by about 4 °C at the Eocene thermal maximum and successively dropped by about 10 °C in the Holocene. This inversion of climatic conditions resulted in the appearance of polar ice caps, first in Antarctica and successively in the Arctic, and the alternation of ice expansion and retreat during the Quaternary.

Climatic variations during the Cenozoic have been attributed to various factors. The large volcanic eruptions, which commenced at about 62 million years with the opening of the North Atlantic Ocean as North America separated from Eurasia, poured enormous amounts of CO_2 in the atmosphere, driving the increase in temperature that led to the Eocene thermal maximum. Successively, the Indian continent began to impact Eurasia about 50 million years ago, activating the uplift of the Himalaya Mountain Range which exposed large volumes of rocks to weathering at the Earth's surface. Atmosphere-rock interaction resulted in the consumption of large quantities of atmospheric CO_2, which triggered an inversion of the thermal warming trend toward cooler climatic conditions. Finally, continental drift modified

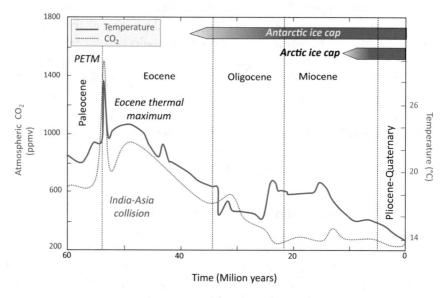

Fig. 8.3 Variation of atmospheric carbon dioxide and temperature during the Cenozoic[9]

---

[9] Zachos et al. [6], Beerling and Royer [7].

the global oceanic circulation, isolating some large regions, such as Antarctica, on which a thick ice cap began to form in the Oligocene.

In the general trend described above, there were some rapid warming events—referred to as **hyperthermals**—often coupled with an increase in atmospheric CO_2. The best documented occurrence—known as the **Palaeocene–Eocene Thermal Maximum** (**PETM**)—took place about 55 million years ago.[10] Such an anomaly was discovered at the end of the last century when abnormal oscillations of oxygen and carbon isotope compositions were detected in shells of Upper Palaeocene benthic (i.e. bottom-dwelling) foraminifera recovered by drilling of seafloor sediments.[11]

Rapid jumps of temperature by about 5–6 °C, with peaks of 8 °C at poles, have been deduced from oxygen isotope studies of foraminiferal shells (see Box 6.1). Moreover, carbon isotope data suggest that some thousand gigatons of carbon dioxide were released into the atmosphere and oceans. The influx rate of carbon is estimated at 0.5–0.6 gigatons per year, which is an order of magnitude less than the present-day anthropogenically driven carbon flux into the atmosphere—a most compelling evidence that human activities can force the Earth's ecosystems much more severely than some natural phenomena. This hot spell lasted about 20,000 years but the exogenic Earth system took about 100,000 years to return to pre-PETM conditions because of the slow consumption of CO_2 by chemical weathering of silicates. The source(s) of massive carbon injection is uncertain. The most popular hypothesis is that CO_2 was released from sediments rich in carbon compounds, which were intruded by large lava bodies during the opening of the North Atlantic Ocean. Carbon delivered by the impact of an extra-terrestrial body has also been postulated.

8.3.3.4 Pliocene to Pleistocene (5.33 to 0.0117 million years)

During the last 5 million years (Pliocene–Pleistocene-Holocene), the Earth's climate has been characterised by short-period oscillations superimposed upon glacial–interglacial alternations occurring with a 100,000 years periodicity.

---

[10] McInerney and Wing [8].

[11] Foraminifera are single-celled microorganisms particularly suitable for paleoenvironmental studies because past fluctuations in physical–chemical conditions of seawater leave a clear record in the oxygen and carbon isotope compositions of their carbonate shell (Box 8.1). Kennett and Stott [9].

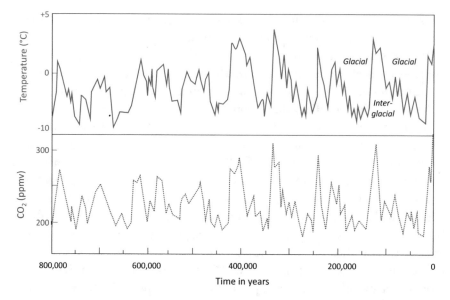

Fig. 8.4 Changes in atmospheric CO_2 concentration and temperature during the last 800,000 years as deduced from studies of the Vostok Station and EPICA ice cores. Carbon dioxide has been measured in the air bubbles entrapped in the ice; temperature is calculated from the isotopic composition of ice (Box 8.1)

Temperature and CO_2 oscillations are particularly well documented for the last 800,000 years by isotopic and chemical studies of the ice samples recovered from drilling at the Russian **Vostok Station** and **EPICA (European Project for Ice Coring in Antarctica)** sites. Stable isotope data on ice samples and geochemical analysis of entrapped air bubbles highlight striking synchronous changes of ambient temperature and CO_2, with carbon dioxide levels in the atmosphere rising from about 180 ppmv to about 260–280 ppmv from glacial to interglacial episode and parallel temperature oscillations of about 10 °C (Fig. 8.4).

8.3.3.5 The Holocene (11,700 years to present)

The latest large climatic change occurred at the transition from Late Pleistocene to Holocene, between about 20,000 and 11,700 years ago, with the shift from the last ice age to the present-day interglacial phase. Over this transition, atmospheric CO_2 concentration increased from about 200 to 260 ppmv, temperature increased by approximately 3–4 °C, the glacial ice cover of the Northern Hemisphere melted extensively and retreated to polar latitudes, and sea level rose about 130 m.

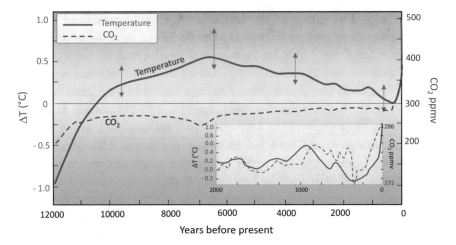

Fig. 8.5 Changes in atmospheric CO_2 and temperature in the Holocene. Variation of temperature ΔT reported on the left ordinates is the difference with the average of the last decades of the twentieth century. The scale of CO_2 is shown on the right ordinates. The inset is a detail of the last 2000 years registered at the Law Dome, Eastern Antarctica[12]

Climatic conditions remained stable throughout the Holocene (Fig. 8.5). According to currently accepted models based on multi-proxy data,[13] there have been modest temperature fluctuations (less than one degree Celsius) with rare thermal anomalies such as the sudden drop occurred in 536 A.D. —probably related to a large volcanic eruption—and minor CO_2 increase at 8000–7000 years, likely related to the agricultural activities.[14]

A widely acknowledged relatively cool episode known as the **Little Ice Age** occurred in the Northern Hemisphere between the end of the Middle Age and the first half of the 19th century. A temperature drop of about 1 °C was accompanied by a modest decrease in atmospheric CO_2 concentration (Fig. 8.5, inset). Such a relationship is highlighted by studies of ice cores from Greenland, and has been also observed by Australian glaciologists in the Law Dome, Antarctica, probably indicating that a climatic variation did

---

[12] MacFarling Meure et al. [10].

[13] Kaufman et al. [11]

[14] However, it has been recently suggested that a monotonic warming occurred through the entire Holocene, in response first to retreating ice sheets and then to the rising greenhouse gas concentrations related to human activities. According to these models, previous global reconstructions reflect the evolution of seasonal, rather than annual temperature.

Bova et al. [12].

occur globally. The causes of the Little Ice Age could be related either to repeated volcanic eruptions (Chap. 3) or to decreased solar activity.[15]

The exit from the Little Ice Age started at the end of the nineteenth century, with an unprecedented rapid rise of both temperature and atmospheric CO_2. Only the global climate change produced by meteorite impact at the end of the Cretaceous was more rapid than that occurring at present. According to researchers at the University of Potsdam, the cooling trend of the Little Ice Age could have led to a new glaciation. However, such an outcome was hindered by the relatively high CO_2 concentrations produced by agricultural activities and the beginning of the industrial revolution. The main conclusions of this study are that global cooling can occur for various reasons, but widespread glaciations can only start at very low concentrations of atmospheric CO_2 (less than about 240 ppmv). The implication is that the relatively high amounts of greenhouse gas generated over the last two centuries have protected us from a new glaciation and will continue to do so for at least the next 50,000–100,000 years. This is an example of just how complicated the Earth system is!

8.3.4   A Lesson for the Future

The account of the climatic history of the Earth reported above, albeit brief, incomplete, and based on models and data that are potentially subjected to revision, provides a wealth of information from which a few conclusions can be drawn. First-order evidence shows that warmer climatic conditions than at present have occurred at different times of Earth history. These **hothouse** states have been interrupted by starkly cooler climatic intervals, referred to as **icehouse** events, during which ice caps formed at poles and in a few extreme cases covered a large part of the planet. The present time is characterised by polar ice caps and, therefore, can be considered as an icehouse state. Intervals of global glaciation can be long-lived, although the transition to icehouse states can be rather short, on the order of a few thousand years.

Another important piece of information is that climatic oscillations are related to various factors, external to the intrinsic spheres of the Earth—atmosphere, ocean, vegetation, rocks, and soil—whose interaction regulates the global climate. The long-term variations are generally related to fundamental geological processes, such as changes in the positions of the continental masses and the formation of mountain belts. Medium-term fluctuations,

---

[15] The Little Ice Age was partially coeval with low solar activity characterised by a small number of sunspots referred to as the Maunder minimum.

lasting tens of thousands to a hundred thousand years, are related to recurrent modification of orbital parameters of the planet, such as the eccentricity of its orbit around the sun, its axial tilt, and precession of the equinoxes. These variations are referred to as **Milankovitch cycles** from the Serbian geophysicist Milutin Milankovitch (1879–1958), who first modelled their effects about a century ago. Short-term factors include volcanic eruptions, the fall of meteorites, and transitory changes of solar activity.[16] However, sometimes, short-lived factors generate self-accelerating runaway effects that can lead to severe long-term climatic scenarios.

An additional relevant message from geology is that, in most cases, both short- and long-lived climatic variations have been accompanied by changes in atmospheric CO_2, which has almost invariably increased during hot period and decreased during cool climatic phases. In short, there has been a close temporal relationship between temperature variations and variations in atmospheric CO_2 concentrations. Such a relationship is particularly significant for the last 800,000 years.

The positive correlation between temperature and CO_2 is not a proof of causal relationship, and both may be independent effects of different causes. For instance, some suggest that the decrease of CO_2 during the Pleistocene glacial periods was related to the arrival into the sea of high amounts of nutrients, especially iron-rich dust particles. Dust particles were collected by winds in the sparsely vegetated or nearly desert continents that dominated the Earth during glaciation and brought to the ocean where iron nourished the marine phytoplankton; arrival of extra iron to the sea promoted a phytoplankton explosion that considerably improved the efficiency of the biological pump and, consequently, increased the consumption of carbon dioxide[17], In such a view, the cause of glaciation was related to some external factor and the reduction in CO_2 was subsequent and actually an effect of cooling.

In some cases, high levels of atmospheric CO_2 preceded and most likely triggered thermal anomalies, supporting the idea that CO_2 is a primary driver of Earth climate.[18] Such a conclusion seems reasonable since CO_2 is a very effective greenhouse gas and its increase, whatever the cause, can well elevate tropospheric temperatures. Such a conclusion is particularly compelling for the present time, which is witnessing an unprecedented rapid increase of both CO_2 and temperature, far above any natural event during the Holocene, and possibly the entire Phanerozoic. The decline of solar activity since 1990 seems

---

[16] LeMouël et al. [13].

[17] Martin [14]; Stoll [15]

[18] Broecker [16].

to exclude external forcing on climate, thus indirectly supporting a role of CO_2 as a major driver of the present global temperature anomaly.

Going to the near future, it is calculated that, if business continues as usual, humans will release several thousand gigatons of CO_2 into the atmosphere during the next hundred years. Some of this amount will be absorbed by other reservoirs of the short cycle, such as vegetation and seawater. However, about 60–70% will remain in the atmosphere, bringing CO_2 concentration to 800–900 ppmv by the end of this century and 2000 ppmv by the year 2400. This range of values is equal to or higher than that of the Palaeocene–Eocene thermal maximum when temperatures were about 10–12 °C higher than at present, and polar ice caps were lacking (Fig. 8.3). In short, man can prompt in a few years the atmospheric and climatic changes that the Earth has generated over millions of years. This is indeed a serious reason for concern that should urge severe actions to reduce the production of CO_2, remove it from the atmosphere, and store this CO_2 in safe places through geological sequestration. Human beings cannot do much to avoid most of the processes that fuel climate modification on Earth. However, they can do a lot to keep under control at least one of these factors, namely the production and handling of greenhouse gases, which have indeed the potential to induce severe climatic change. These actions need to follow geological criteria to be efficacious,[19] as discussed in Box 8.2.

8.4   Plastic Waste Pollution

The term *plastic* is used to indicate various materials, including numerous synthetic polymers—such as polyvinyl chloride (PVC), polyethylene (PE), polyethylene terephthalate (PET). Polymers are long molecules that are very common in nature (e.g. the cellulose from which paper is made), but can also be produced synthetically. About 350–400 million metric tons of plastic polymers are produced annually worldwide. Much of the plastics material accumulates as waste in the environment. Its extremely slow biodegradation rate causes survival for years to decades or longer. Plastic and chemical substances added during manufacturing to acquire certain performance properties can alter the way hormones normally work in animal bodies, favour cancer, cause birth defects, and damage the immune system.

Plastic objects are widely dispersed worldwide, even in the most remote places, such as the slopes of Mount Everest or the deep ocean (see the special

---

[19] Kharaka and Cole [17].

issue: *Our plastic dilemma*. Science, 2021, Vol. 373, Issue 6550). In an ocean environment, plastic waste can sink, float, or disperse as microparticles in the water; micro-plastic fragments can be ingested by marine animals as small as plankton, thus entering the food chain and inevitably reaching humans.

Floating plastic trash does not distribute randomly, but is gathered by the currents and winds and concentrated in layers up to several meters thick, which can extend over the sea surface for thousands of square kilometres. Marine plastic accumulations are called 'garbage patches' and contain all kinds of materials, including fishing nets, bottles, bags, and bins; they are also home to algae, plankton, and an exceedingly large number of bacteria and viruses. The largest of these 'islands' covers a wide area of the ocean surface, located approximately between California and Hawaii.

Recent studies highlight that plastic is also invading the atmosphere in the form of micrometres to a few millimetres sized particles. Plastic objects released into terrestrial and marine environments break down into small bits that are entrained by winds and dispersed worldwide. As a result, air-born microplastic fragments are widely dispersed both in populated and remote areas, including the Pyrenees, Alps and the Arctic. It has been estimated that more than one thousand metric tons of plastic microfibers are floating in the air of the Western USA, including the protected areas, such as Yellowstone Park.[20]

The largest fraction of the microplastic particles and fibres come from crowded roads and urban centres. Another 10–15% comes from the breakdown of mismanaged waste, including those dispersed in the oceans. To put it plainly, so much plastic has accumulated in the ocean that part of it is returned as microparticles to the land.

The microplastic in the exogenic environment is creating an independent sort of geochemical cycle akin to global biogeochemical cycles of natural elements. Since plastic polymers are synthesised from petroleum, the dispersion of plastics is an additional way of transferring carbon from deep reservoirs to the external environment. Therefore, the plastic cycle in the environment is an aspect of the anthropogenically driven alteration of the carbon cycle.

Since the amount of plastic production is increasing by about 5% per year, it is expected that billions of metric tons of plastic will be accumulated in the coming years. This makes plastic dispersion in the terrestrial ecosystems an increasingly severe problem that is worthy of consideration and concern.

---

[20] Brahney et al. [18].

8.5 Summary

The geochemical cycles of elements have been operating through the entire history of the Earth, following slow but continuous processes, and sometimes undergoing dramatic modifications by catastrophic events.

During the Holocene, the natural geochemical cycles of elements have suffered substantial alteration because of anthropic activities. Elements have been subtracted from their natural cycles and forcefully introduced into the atmosphere, hydrosphere, and biosphere. Such a transfer sped up dramatically during the last century.

Anomalously high amounts of many elements that are harmful to humans and other animals, such as lead, mercury, and arsenic, have been spread through in the environment, becoming a dramatic health problem because of heavily polluted soils, atmosphere, and waters.

A particularly debated issue is the upsetting of the carbon cycle. This element has been transferred in large quantities from the lithosphere to the atmosphere, due to the burning of fossil fuel and the calcination of lime-stone for cement production. This massive redistribution of carbon has led to an increase in CO_2 in the atmosphere and oceans, with disturbing consequences for the Earth's climate, water chemistry, and marine biota. The high rate at which these processes are taking place is particularly worrying, being unprecedented in the geological record.

The widespread dispersion of plastic, both on the continents and in the oceans, is another aspect of the disruption of the carbon cycle by humans. The increasing amount of microplastic in the ocean and air is endangering both the marine and terrestrial biosphere.

The effects of human activities on the environment are a fundamental concern of modern science. Any geologically sound solution must respect the natural geochemical cycles and should be aimed at bringing the elements from the external environment back to their original geochemical reservoirs (Box 8.2).

8.6 Box 8.1—Stable Isotopes: Measuring the Temperature of the Past

Paleoclimatologists rely upon a wide variety of geological information to reconstruct the past climate of the Earth. Climatic indicators in the geological record include some spectacular glacial erosional features, sedimentary rock types such as the glacial deposits known as **tillite**, the distribution of fossil forms of life, and some kinds of geochemical data. Temperatures prevailing in the oceans and at the Earth's surface during the geological past (**paleotemperature**) can be deduced from the contents of certain stable isotopes in fossil shells and other geological materials (e.g. the terrestrial carbonates from cave and paleosols) whose isotopic signatures depend on the ambient temperature during their formation. Understanding how such 'geothermometres' work requires some basic knowledge of the elemental principles of isotope geochemistry.

8.6.1 Some Basics of Isotope Geochemistry

Each chemical element is characterised by a specific number of protons inside the nucleus (**atomic number** or Z), but may contain a variable number of neutrons, thereby exhibiting different mass (M = sum of protons plus neutrons). The atoms of a given element, showing the same Z but a variable number of neutrons in their nuclei, are called **isotopes**. For instance, the element oxygen (O) is characterised by a specific atomic number $Z = 8$, but atoms occurring in nature have a variable number of protons, from 8 to 10, and, therefore, a mass number M = 16 to 18.

An isotope of an element is indicated by the symbol of the element and the mass number reported as a superscript on the left-hand side of the symbol; for instance, ^2H, ^{16}O, ^{18}O, ^{36}S, ^{204}Pb, ^{238}U denote the hydrogen, oxygen, strontium, lead and uranium isotopes with their mass numbers (i.e. 2, 16,18, 36, 204, 238). Isotopes can also be specified by giving the name of the element and the mass, separated by a hyphen (e.g. helium-4, oxygen-18, uranium-238).

All the isotopes of a given element have the same external electronic structure and, therefore, have the same chemical behaviour. However, the different weights make the various isotopes move at a different speed during chemical and physical processes.

Isotopes can be stable or radioactive. The radioactive isotopes change their nature with time because of the spontaneous transformation into other isotopes. This process—known as **radioactive decay**—has a paramount geological interest because it is a primary source of heat to the Earth and allows for the determination of the absolute ages of minerals and rocks (Box 9.1). By contrast, the relative abundances of stable isotopes (isotope ratios) in any given substance remain constant through time. However, their abundance can change during physical and chemical processes, a modification that is known as **isotope fractionation**. Only isotopes of light elements, with M < 35–40, can fractionate significantly during chemical and physical processes because their relative mass differences are high and their reaction kinetics vary significantly as a function of the isotopic mass.

The basic principle of using stable isotopes of light elements as geother-mometers is that the extent of isotope fractionation is a function of the temperature at which the chemical and physical processes take place. There-fore, measuring of isotope compositions of materials allows deducing the ambient temperature of their formation. The application of stable isotope fractionation to geothermometry was first developed by the *Nobel Laureate* Harold C. Urey (1893–1981) of Chicago University and successively devel-oped by many others, especially the Italian-American geologist Cesare Emiliani (1922–1995), a founder of oceanography.

8.6.2   The *Delta* Notation

The isotopic composition of geological materials, or any other substance, is not expressed as the absolute concentration of single isotopes—as done for major and trace elements—but as ratios of two isotopes, in which the most abundant is placed by convention as the denominator. For instance, the hydrogen and oxygen isotope composition of geological materials are expressed by the ratios $^2H/^1H$ and $^{18}O/^{16}O$. Ratios of isotopes are more easily determined by analytical techniques than single isotope abundances; at the same time, they allow a better perception of the relative enrichments of light versus heavy isotopes in geological samples or any other material during chemical or physical transformations.

Values of isotope ratios in natural samples are generally very low, and their variability is in the range of the fourth or fifth decimal place. Since these numbers are difficult to handle, they are transformed to more friendly figures by dividing (normalising) the isotope ratio of any substance against the value of a reference standard and multiplying the result by one thousand. This simple mathematical manipulation yields the so-called *delta per mil* (δ‰) value,[21] consisting of one to three integer figures, easy to remember, and simple to plot on a graph or make calculations.

Various reference standards are used for different isotopes. For instance, the mean ocean water (SMOW) is adopted as the normalising standard of hydrogen and oxygen isotope ratio.

Delta values can be negative or positive, depending on whether the isotope ratio of the analysed sample is lower or higher than the reference standard.

---

[21] The formula to calculate the *delta per mil* is: $[(R_{sample}-R_{standard})/R_{standard}$ x; $1000]$ where R_{sample} and $R_{standard}$ are the isotope ratio measured in the sample and in the reference standard, respectively.

Samples with low *delta* values are preferentially enriched in light isotopes than samples with a high *delta*.

8.6.3 Oxygen-Hydrogen Isotope Geothermometry

Hydrogen and oxygen isotope compositions of natural substances (e.g. water) are the most commonly used as geothermometers.

Hydrogen comprises two stable isotopes, denoted as ^1H and ^2H and respectively named *protium* and *deuterium* (D). The latter contains one proton and one neutron in its nucleus; protium has only one proton and no neutrons. Protium weights one half of deuterium and represents some 98% of natural hydrogen.

Oxygen has three stable isotopes, ^{16}O, ^{17}O and ^{18}O, with the heaviest ^{18}O weighing 12.5% more than the lightest ^{16}O. The latter is much more abundant in nature (99.76%) than ^{17}O and ^{18}O.

Because of the strong relative mass difference, isotope compositions of hydrogen and oxygen change significantly during chemical-physical processes, such as water evaporation or condensation or mineral crystallisation. The amount of isotope fractionation is a function of the temperature at which the processes take place. Therefore, hydrogen and oxygen isotopes can furnish precious thermometric information.

Let's consider natural water. Its molecules contain various combinations of hydrogen and oxygen isotopes. The lightest molecules are those formed by the two atoms of light hydrogen (^1H) and one atom of light oxygen (^{16}O): its formula can be written as $^1H_2{}^{16}O$ (molecular mass = 18). The heaviest molecules are formed by heavy isotopes of the two elements and can be written as $^2H_2{}^{18}O$ (molecular mass = 22). Heavy and light molecules have identical chemical behaviour but widely different reaction kinetics, i.e. they move at different speed during chemical and physical processes.

Consider the case of evaporation. The lighter molecules $^1H_2{}^{16}O$ pass to the vapour phase more quickly than the heavy ones $^2H_2{}^{18}O$. As a result, the water vapour is more enriched in light oxygen and hydrogen than in the starting liquid water. The opposite occurs during precipitation, and rainwater is heavier than atmospheric moisture.

A key aspect is that the proportions of light *versus* heavy molecules passing into the vapour phase (i.e. the ratios between heavy and light isotopes, or the isotope fractionation) increase with decreasing temperature. In other words, the evaporation of light molecules is more strongly favoured in the cold regions of the Earth than at hot temperatures. The consequence is that the water vapour—and hence, precipitations—at high latitudes is more strongly

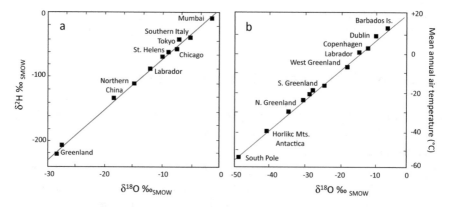

Fig. 8.6 (a) Variation of oxygen and hydrogen isotope compositions in meteoric water from localities sited at various latitudes; (b) Plot of the oxygen isotope composition of precipitations against mean annual temperatures for representative worldwide sites

enriched in light molecules than vapour at the tropics or the equator. In short, δ^2H‰ and $\delta^{18}O$‰ of precipitations at poles are lower than in the tropics and equator.[22]

The relationship between H and O isotope compositions and temperature is highlighted by the very significant positive correlation of δ^2H‰ *versus* $\delta^{18}O$‰ for precipitations worldwide (Fig. 8.6a). A similar relationship is shown by $\delta^{18}O$‰ *versus* mean annual temperatures for various regions at a global scale (Fig. 8.6b). Simple equations allow calculating precise values of temperatures, starting from the measured isotopic composition of precipitations.

It has been previously observed that ice drilling in Antarctica revealed oscillating values of atmospheric CO_2 and ambient temperature during the last 800,000 years (Fig. 8.4). Carbon dioxide was directly measured in air bubbles trapped in the ice. Temperatures were calculated from the oxygen and hydrogen isotope composition of the ice collected at various depths along the drilling.

Palaeotemperatures can also be deduced from $\delta^{18}O$‰ in carbonate fossil shells, a field of research pioneered and developed by Cesare Emiliani. A main outcome of these studies has been the reconstruction of oxygen isotopic

---

[22] Although correct, things are a bit more complicated than described in this paragraph. As a matter of facts, little to no evaporation occurs at high latitudes. Therefore, isotope fractionation is related more to the global water cycle than to local processes. In short, the water evaporated at low latitudes is transported poleward as atmospheric moisture and progressively condenses and precipitates, losing heavy molecules. The remaining vapour is gradually depleted in heavy molecules, and precipitation results in the very low *delta* values of snow and ice at high latitudes.

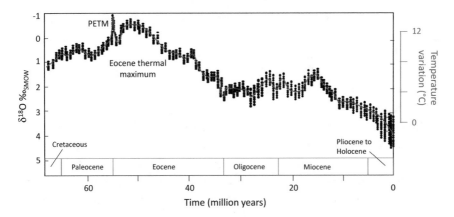

Fig. 8.7 Variation of δ ^{18}O composition (‰ SMOW) of benthic foraminifera during the Cenozoic era. Isotopic compositions reveal a temperature drop of some 12–14 °C from the Eocene thermal maximum to Holocene

variations of foraminiferal shells collected from ocean floor sediments. Such variations have allowed tracing quite accurately the temperature variation during the entire Cenozoic era. The profile is shown in Fig. 8.7.

Stable isotope geochemistry has many other applications, not only in geology but also in biology, astronomy, and other fields of natural sciences. For instance, fractionation of light isotopes, such as carbon, occurs during biological processes such as photosynthesis. The isotope fractionation allows tracing the photosynthetic pathways followed by different plants in the various terrestrial and marine environments and the temperature conditions they take place. Carbon and nitrogen compositions in animals undergo significant fractionation as these elements pass up the food chain. Therefore, it is possible to infer the diet of animals from their carbon and nitrogen isotopic composition. Stable isotopes are commonly used to track the pathways of individual elements in the human body and to determine the basic metabolic mechanisms. Finally, the ratios of hydrogen isotopes (δ^2H) in the ice of most comets is markedly different from those found in the Earth's ocean. This has led to the conclusion that comets have not contributed significantly to the water budget of the planet.

8.7 Box 8.2—CO_2 Sequestration: The message from *Decameron*

In the tenth novel of the third day of the *Decameron*—a collection of novellas by the Italian author Giovanni Boccaccio (1313–1375)—Dioneo narrates how the monk Rustico convinced the hermit girl Alibech that the 'most acceptable service that could be rendered to God was to put back the devil into the hell, the natural place whereto he had condemned him'.

Although the aims of father Rustico were far away from ecology, the novel could teach us it is wise that everything is in its own place, a principle that fits perfectly with the environmental issue of geochemical cycles.

There is plenty of evidence that the increase of CO_2 in the atmosphere during the industrial era is linked to the forceful and fast transfer of considerable amounts of carbon from their long-term 'natural place' of residence to the short-cycle reservoirs of the atmosphere and hydrosphere. The Earth's external system is unable to absorb the excess CO_2, which is accumulating in the atmosphere and oceans.

Studies on how to overcome the imbalances caused by human activities are many, as can be seen from the documents of the **Intergovernmental Panel on Climatic Change** (IPCC) available on the web.

The most straightforward solution is to reduce fossil fuel consumption and convert to alternative energy that is collected from renewable sources such as the sun, wind, tides, and geothermal heat. After all, we have more than sufficient energy delivered from the Sun or sitting inside our planet, and it is at odds with common sense that humans concentrate on fossil fuel energy rather than using largely available and renewable sources. Nevertheless, consumption of fossil fuels is constantly increasing and additional solutions are to be sought to address the issue.

A popular proposal is to proceed with extensive forestation that increases the absorption of excess CO_2 through photosynthesis. However, it has been argued that this operation could reduce the albedo, leading to the opposite of the result being sought. Furthermore, recent research has shown that forests, if not well protected, can produce more CO_2 through degradation, fires, and poor management practice than they can absorb through photosynthesis. Finally, this solution does not appear optimal from a geological point of view because it does not bring carbon back to the long cycle from which it has been removed; instead, it simply redistributes carbon between two reservoirs of the short cycle, namely the atmosphere and the biosphere. A large fire or a period of intense drought would cause a rapid return of CO_2 to the atmosphere, thus nullifying such efforts.

However, extensive and well-executed reforestation in the appropriate areas of the planet could contribute effectively to the temporary sequestration of CO_2. According to the IPCC, more than 700 billion tonnes of CO_2 could be removed from the atmosphere by the end of this century through intensive reforestation. Many countries have committed to following the indications of international scientific agencies, but it is hard to say if they will meet their pledges.

Another action proposed is to pump CO_2 deep into the ocean, where it would be transformed into a solid hydrated form ($CO_2.nH_2O$) that could remain on the seafloor for thousands of years. However, it would be like sweeping the dust under the carpet, hoping that nothing will come to disturb the hiding (e.g. a large submarine eruption).

The most geologically correct choice should be to restore natural equilibrium, bringing carbon back into the long cycle reservoirs, which means putting 'back the devil into the hell'.

A geochemically effective process is to make CO_2 react with silicate minerals, such as olivine or pyroxenes, to form serpentine, calcite, and magnesite, as described by the chemical reaction:

$$2Mg_2SiO_4[olivine] + CaMg(SiO_3)_2[diopside] + 3CO_2 + 2H_2O$$

$$\rightarrow Mg_3Si_2O_5(OH)_4\,[serpentine] + CaCO_3[calcite] + 2MgCO_3[magnesite] + 2SiO_2$$

$$(8.1)$$

Minerals produced by these reactions are very stable, last indefinitely over time, and can be considered as reservoirs of the long cycle. Such transformations could be carried out in artificial plants that require energy to speed up the reactions. An interesting alternative may be to inject CO_2 into deep rocks containing olivine and pyroxenes (such as basalts and peridotites), and to let the reactions take place thanks to the heat provided by the normal geothermal gradient. Basalts and peridotites are very common in nature, both along orogenic belts (ophiolite series) and in anorogenic geological settings (e.g. India, Iceland). An encouraging experiment has been undertaken in Iceland, where a mixture of CO_2 and hydrogen sulphide was dissolved in water and injected into basaltic rocks at relatively shallow depths. After a couple of years, the *in-situ* check disclosed that about 95% of CO_2 had been stabilised by the formation of carbonate minerals.[23]

Another solution could be storing CO_2 by injection into deep rock formations, thus returning to the lithosphere the carbon that had been subtracted by extraction of fossil fuels. Such a procedure is complex, costly, and occasionally ineffective. The best sites for storage are porous rocks covered by impermeable layers that may guarantee sealing and potentially prevent the gas from escaping to the surface or interacting with subsurface aquifers. Sites should be located in areas away from dormant or active faults that could

---

[23] Matter et al. [19].

be reactivated by highly pressurised fluids. Exhausted hydrocarbon deposits (e.g. former oil fields) are natural sites potentially suitable for such geological storage and sequestration. The gas capture must take place immediately after its production before it diffuses into the atmosphere; this is possible in large industrial settlements or cement factories, which together emit about 60% of anthropogenic CO_2. Then, pipelines must be built to bring CO_2 to the sequestration sites after purification and liquefaction. At the moment, only twenty or so such plants operate, ensuring the sequestration of around 40 million tonnes of CO_2 per year.

Even more complex is the removal of the CO_2 that has accumulated in the atmosphere. Besides the reforestation mentioned above, some suggest that a significant number of industrial plants should be constructed that take in air, separate the CO_2 through reactions with alkali substances, and produce pure CO_2 for industrial use or geological storage at appropriate sites. The technology is already available and operational, but the costs are still too high. The amounts of CO_2 sequestered annually constitute only a tiny fraction of that produced by fossil fuel consumption.

References

1. Harmon RS, Parker A (2011) Frontiers in geochemistry. Contribution of geochemistry to the study of the Earth. Wiley-Backwell, 263 p
2. Suarez CA, Edmonds M, Jones AP (eds) (2019) Catastrophic perturbations to Earth's deep carbon cycle. Elements 15(5)
3. Allègre C (2007) Ma vérité sur le planet. Plon Fayard, 237 p
4. Lear CH, Anand P, Blenkinsop T, Foster GL, Gagen M, Hoogakker B, Larter RD, Lunt DJ, McCave IN, McClymont E, Pancost RD (2020) Geological Society of London Statement: what the geological record tells us about our present and future climate. J Geol Soc 178:1–13
5. Scotese RC, Song H, Mills BJW, van der Meer DG (2021) Phanerozoic paleotemperatures: the Earth's changing climate during the last 540 million years. Earth-Sci Rev 215:103503
6. Zachos J, Pagani M, Sloan L, Thomas E, Billups K (2001) Trends, rhythms, and aberrations in global climate 65 Ma to present. Science 292:686–693
7. Beerling DJ, Royer DL (2011) Convergent Cenozoic CO_2 history. Nat Geosci 4:418–420
8. McInerney FA, Wing SL (2011) The Paleocene-Eocene thermal maximum: a perturbation of carbon cycle, climate, and biosphere with implications for the future. Ann Rev Earth Planet Sci 39:489–516

9. Kennett JP, Stott LD. (1991) Abrupt deep-sea warming, palaeoceanographic changes and benthic extinctions at the end of the Palaeocene. Nature 353:225–229

10. MacFarling Meure C, Etheridge D, Trudinger C, Steele P, Langenfelds R, Van Ommen T, Smith A, Elkin J (2006) Law dome CO_2, CH_4 and N_2O ice core records extended to 2000 years bp. Geophys Res Lett 33:L14810

11. Kaufman D, McHay N et al (2020) Holocene global mean surface temperature, a multi-method reconstruction approach. Sci Data 7:201

12. Bova S, Rosenthal Y, Liu Z, Godad SP, Yan M (2021) Seasonal origin of the thermal maxima at the Holocene and the last interglacial. Nature 589:548–553

13. LeMouël J-L, Blanter E, Shnirman M, Courtillot V (2009) Evidence for solar forcing in variability of temperatures and pressures in Europe. J Atmos Solar-Terrest Phys. doi:10.1016/j.jastp.2009.05.006

14. Martin JH (1990) Glacial-interglacial CO_2 change: the iron hypothesis. Palaeoceanography 5:1–13

15. Stoll H (2020) Thirty years of the iron hypothesis of ice ages. Nature 578:370–371

16. Broecker W (2018) CO_2: Earth's climatic driver. Geochem Perspect 7:117–196

17. Kharaka YK, Cole DR (2011) Geochemistry of geologic sequestration of carbon dioxide. In: Harmon RS, Parker A (2011) Frontiers in geochemistry. Contribution of geochemistry to the study of the Earth. Wiley-Backwell, pp 133–173

18. Brahney J, Mahowald N, Prank M, Cornwell G, Klimont Z, Matsui H, Prather KA (2021) Constraining the atmospheric limb of the plastic cycle. Proc Natl Acad Sci 118(16)

19. Matter JM, Stute M, Snæbjörnsdottir SÓ, Oelkers EH, Gislason SR, Aradottir ES, Sigfusson B, Gunnarsson I, Sigurdardottir H, Gunnlaugsson E, Axelsson G (2016) Rapid carbon mineralization for permanent disposal of anthropogenic carbon dioxide emissions. Science 352:1312–1314

9

From Hadean to Anthropocene—The Endless Story of a Lucky Planet

Die zwei grössen Tyrannen del Erde: der Zufall und die Zeit (The two grant tyrants of the Earth: Time and Chance)
Johann Gottfried Herder (1744–1803)

9.1 Introduction

The history of the Earth is a major field of study in the Geological Sciences, but all the branches of Natural Sciences, including chemistry, physics, astronomy, and biology have contributed to its knowledge. Several centuries of studies provided an enormous amount of information on such a key topic, although our knowledge is still inadequate, especially for remote eras, the geological record of which is scanty and poorly preserved.

Most of the information on the history of the Earth comes from the study of rocks cropping out on the Earth's surface. The rock record, however, is very discontinuous and almost completely lacking for the first aeon of the Earth history. The oldest known rocks have ages of about 4.0 million years; some minerals date back to 4.4 billion years. There are approximately 500 million years for which there is little or no geologic record, which makes the early history of the Earth basically unknown. Information becomes progressively richer as the time comes closer to us, although most of the geological processes remain hypothetical because evidence has been cancelled by the continuously evolving crust. Therefore, large sectors of the Earth's history are still highly contentious and matter for speculation.

The general picture, however, is clear. The history of the Earth is the result of the interplay of normal physical–chemical processes—such as heat loss and the action of gravitational forces—and a high number of extraordinary events, such as the impact of extraterrestrial bodies, gigantic volcanic eruption, and other catastrophic events. All these phenomena efficaciously interacted to turn an initial hot sphere into a hospitable planet, a unique case in the solar system and the known Universe.

9.2   From the Solar Nebula to the Formation of Planets

The solar system formed by the compaction of a cold molecular nebula about 4600 million years ago, over nine billion years after the birth of the Universe. Hydrogen and other components assembled in a gravitational centre, causing a local increase in density and acceleration in the rotation speed of the nebula, which flattened to take the shape of a disk. Inside the gravitational core, the pressures and temperatures built up progressively, finally reaching the conditions necessary to activate nuclear fusion of hydrogen and create the Sun.

The solar disk had strong internal thermal and compositional gradients. Volatile substances such as hydrogen, helium, nitrogen, and water concentrated in the cooler outer zones, ultimately forming the great gaseous planets, Jupiter, Saturn, Uranus, and Neptune.

In the internal regions, the heavier chemical elements and compounds were in a gaseous state due to very high temperatures. Gradual cooling led to condensation, generating a sort of fine dust, which successively aggregated into larger and larger stones and then kilometre-sized bodies or **planetesimals**. These planetesimals acted as gravitational centres, which attracted additional fragments and dust from their orbits to form the planets of Mercury, Venus, Earth, and Mars.

The opposite process of collision and fragmentation also accompanied accretion. When the latter prevailed, larger planets aggregated. When the disintegration was dominant, such as in the orbit between Mars and Jupiter, the material remained fragmented as irregularly shaped rocky bodies of various sizes.

This complex sequence of events took a relatively short amount of time. The formation of the solar disk was completed in a few hundred thousand years; the initial and intermediate phases of growth of planetesimals and planets came to an end after a few million years. The last stages of planet accretion took longer, about 100–150 million years.

The birth date of our planet is set by geochronologists at 4567 million years (Ma) ago.[1] This is the age of some components of a meteorite that fell in the village of Pueblito de Allende, Mexico, in February 1969. The Allende meteorite is the oldest extra-terrestrial object ever found, and its age is adopted, a bit arbitrarily, as the birthdate of our planet.

9.3 Hadean: The Hellish Aeon (~ 4600–4000 Ma)

The time interval (eon or aeon) extending from the Earth's aggregation to 4.0 billion years ago is called **Hadean**. The term comes from the Greek god Ἀιδης (Aides), the king of the underworld. As it will be clear soon, no other name could be more appropriate.

In the early stages of its life, the Earth was an incandescent sphere, wrapped in an ocean of magma and surrounded by a burning primordial atmosphere. The heat necessary to keep the Earth burning was provided by the decay of radioactive isotopes, whose abundances were much higher than at present. The kinetic energy released by the compaction of the planet under its own weight and continuous meteoritic bombardment contributed strongly to the thermal budget.

Quite 'soon', heavy metals, such as iron and nickel, separated from the silicate mass and sank towards the centre to form a core (Chap. 3, Fig. 3.3). This differentiation took a few tens of millions of years to complete.

Conditions were gradually becoming less extreme when a cataclysmic cosmic collision radically altered the course of events. About 40–50 million years after the beginning of the Earth aggregation, when the protoplanet had reached about 90% of its mass and the core had partly separated, a cosmic body the size of Mars (called **Theia**, the mother goddess of Selene) impacted into the Earth, undergoing disintegration and throwing enormous quantities of material into space. Some of the debris dispersed into open space, but a portion remained as a cloud within the Earth's gravitational field, and successively aggregated to form the Moon.

The Moon is an unusual object in our solar system, as it has a very high mass compared to the planet around which it orbits. This is a fortunate circumstance because the considerable mass of its satellite stabilises the Earth's

---

[1] Allégre et al. [1].

rotational axis. This condition favours fairly uniform irradiation of the Earth by the Sun and establishes a relatively homogeneous temperature distribution across the Earth's surface.

The impact of Theia had crucial consequences for the evolution of the Earth. Collision re-melted its outer layers extensively, gave the planet its axial tilt, and sped up its rotational speed. The short duration of the day on Earth is the effect of this acceleration. A day on sister planet Venus lasts 243 Earth days, with evident consequences for the distribution of solar heat on the planet's surface.

Not much after the collision, a thin rocky crust formed because of solidification at the magma ocean surface. The mantle then began to crystallise, losing gaseous substances such as nitrogen and water, which rose to the surface to form a primitive atmosphere. However, the mantle continued with vigorous convection, cracking and re-cracking the thin crust. The crustal fragments started jostling on the surface of the Earth, setting up an early phase of plate tectonics.

When the external temperatures dropped, water vapour condensed in the atmosphere and could fall back to the surface as rain. There is no conclusive evidence on the time this event started. However, some important clues are provided by the study of very tiny, less than a millimetre-sized zircon crystals recovered from the metamorphic rocks of Jack Hills, Western Australia. Radiometric dating of these zircons revealed ages of 4.4 to 4.2 billion years, much older than their host rocks that are 3.0 billion years old. Such an age discrepancy suggests zircons were formed earlier than the host rocks by crystallisation inside a granite magma body during the Hadean. Granites were successively exposed at the surface and subjected to disruption and erosion. The component minerals were transported away from the source and accumulated in a sedimentary basin where they underwent burial and metamorphism. The incredible aspect of this story is that the zircon crystals remained unaltered during this geological odyssey, keeping their original compositions intact. Studies on these very tiny crystals, therefore, have furnished unique information on the age of crystallisation in the original granite magma and on the geological environment during the Hadean.

A study of the oxygen isotope composition of the Jack Hills zircons revealed that the granite magma had been contaminated by liquid water under surface, or near-surface, conditions.[2] Therefore, water was present on the Earth's surface at the time of granite magma crystallisation. This water must have infiltrated deep into the early crust and mixed with granite

---

[2] Mojzsis et al. [2]

magma, to modify the oxygen isotopic composition of crystallising minerals, including zircon. In conclusion, zircons from Jack Hills demonstrate that water condensed quite early after Earth's final stages of accretion. Zircons also reveal that granitic rocks existed during the Hadean; since granites are the main constituents of the continental crust, it is concluded that the first proto-continental masses started growing around 4.3–4.4 billion years ago.

Overall, the vast storehouse of information provided by zircons is a stunning example of how sub-millimetre-sized objects can furnish paramount information on processes that affect the planet on a global scale. From these and other related studies, the Hadean Earth appears as a planet already structured into core, mantle and crust; the latter had a basaltic composition and was continuously pierced by magmatic blobs rising from depth. The atmosphere comprised nitrogen, water, methane, and ammonia. CO_2 content was initially relatively low, but increased considerably with time because of intense volcanic activity. Oxygen was absent. Water accumulated slowly on the surface until an immense oceanic mass enveloped the planet. The ocean water was hot, acidic, and contained halides (Cl, F), sulphides, and various metals in solution, making the Earth a hostile stifling inferno.

The fundamental question regarding this scenario concerns the possibility of liquid water existing on a planet that was enveloped in a potentially very hot atmosphere containing large quantities of greenhouse gases. The explanation for this is that the greenhouse effect was counterbalanced by a much lower brightness of the Sun that emitted some 30% less energy than at present. Such a phenomenon is referred to as the **Faint Young Sun Paradox**.

9.4 Archaean (4000–2500 Ma): The Dawn of Life and the Start of Modern Plate Tectonics

The transition between the Hadean and Archaean aeons is marked by the so-called **Late Heavy Bombardment**, a shower of meteorites of all sizes that struck the inner planets of the solar system between about 4100 and 3800 Ma ago. The Earth preserves no trace of these impacts because they have been entirely erased by erosion and other dynamic geological processes. However, meteorite impact craters from the Late Heavy Bombardment are well preserved on the lunar surface, where they had remained undisturbed before the astronauts of the Apollo lunar missions picked up rock samples and brought them back to Earth between 1969–1972 for age dating and other types of analysis.

The most significant geological message from the Archaean rocks is the earliest evidence for life found in samples from Canada and Greenland. It is still unknown when and how life began, and if it is entirely terrestrial in origin or some organic molecules, if not tiny living organisms, were carried onto Earth by extra-terrestrial bodies. As a matter of fact, a particular class of meteorites (**carbonaceous chondrites**), including the famous and much-studied Murchison meteorite that fell in 1969 in southeastern Australia, contain several types of organic compounds such as amino acids, carboxylic acids, and hydrocarbons that have an obvious cosmic origin and could have carried the base components of life on the Earth.[3]

Here on Earth, primordial microbial organisms first appeared in marshy areas, where water was intermittently present. The atmosphere provided the chemical components required to form organic compounds, and sunlight furnished the energy for reactions. An alternative place could have been hydrothermal vents on the seafloor, where primitive organisms (**archaea**) continue to live even today. Hydrothermal waters were warm and capable of providing the energy to drive early metabolic cycles that finally resulted in the construction of RNA.

Geological evidence of early forms of life is scanty and this remains a topic of controversy and lively debate. Recently, some 3770 Ma old rocks from Quebec, Canada, were found to contain structures similar to those created by the present-day microorganisms around degassing vents on the seafloor. Other coeval biogenic forms have been observed in rocks from Greenland.[4]

Stromatolites make up the main Precambrian fossil record. These are decimetre sized and roughly cylindrical or conical structures that consist of superimposed thin layers of carbonate, whose origin is mostly linked to the biological activity of **cyanobacteria** or green–blue algae. Cyanobacteria have the precious ability to perform photosynthesis, whereby oxygen and carbohydrates are formed by the reaction between water and carbon dioxide under the effect of sunlight. Stromatolites can still be observed today in various places worldwide, including the Shark Bay, in Western Australia. The first stromatolites probably date back to about 3.7 billion years ago; their wide diffusion during the Late Archaean generated increasing amounts of molecular oxygen, finally causing a radical change in atmospheric composition.

---

[3] The hypothesis of a cosmic origin of life and its dissemination, with different outcomes, on various planets is known as **panspermia**.

[4] Nutman et al. [3].

   Dodd et al. [4].

A geodynamic innovation took place during the Lower-Middle Archaean between about 3.5 and 3.2 billion years ago with the inception of modern-style plate tectonics, although an early form of plate tectonics initiated shortly after the Hadean by about 3.9 billion years.[5] Plate tectonics was an outcome of the progressive cooling of the planet that brought about the hardening of the lithosphere and its fracturing into mobile plates. Subduction of these lithospheric plates dramatically modified the water cycle that affected deep layers of the Earth's mantle and increased its degree of oxidation. As a result, volcanic gases became progressively depleted in reducing compounds (e.g. H_2S, CH_4, SO) and abundant andesitic magmas formed and erupted (Chap. 3). Andesites have a lower content of ferrous iron (Fe^{2+}) than basalts and, therefore, require less oxygen for their oxidation. Overall, the tectonically induced modification of the redox state of the mantle—and hence of magmas and volcanic gases—could have brought about lower consumption of atmospheric oxygen, favouring, together with other processes, its accumulation in the atmosphere and finally leading to the so-called **Great Oxidation Event** to be discussed below.

The emplacement of andesitic magma into the crust contributed greatly to the growth of continents, another fundamental step in the evolution of the Earth.[6] The first supercontinent *Kenorland* probably aggregated during the Neoarchaean (2.7 billion years ago) by merging of various minor blocks that are presently regions of North America and Greenland (*Laurentia*), the Baltic region (*Baltica*), southern Africa (*Kalahari*) and Western Australia. Exposure of progressively larger rock masses at the surface fostered the flow of nutrient elements into the sea, promoting an expansion of early life. At the same time, the interaction between the atmosphere and silicate minerals intensified the consumption of CO_2 through weathering (Chap. 2), while the friction of the ocean tides against the continents gradually slowed down the rotation speed of the Earth, resulting in an increase of the length of a day by about 10–15% or perhaps much more. Notably, the increased exposure to sunlight could have favoured the development of photosynthetic organisms, thus favouring accumulation of molecular oxygen in the Earth's atmosphere and oceans (see Chap. 7).

---

[5] Condie and Pease [5].
[6] Taylor [6]
   Dhuime et al. [7].

9.5 Proterozoic (2500–541 Ma): Oxygen, Eukaryotes, Supercontinents

The transition from Archaean to Proterozoic is set at 2500 Ma, close to the beginning of another major event in the Earth history—the substantial increase in the concentration of atmospheric oxygen known as the **Great Oxidation Event** (**GOE**) that occurred between about 2450 and 2050 Ma ago. During this time, the concentration of atmospheric O_2 rose from a few parts per million up to some 2–3%, i.e. 20,000–30,000 ppm (Fig. 9.1).

Accumulation of molecular oxygen in the atmosphere-hydrosphere system began when the production rate of oxygen by photosynthesis and photolysis exceeded its consumption by reductants such as ferrous iron, some sulphur compounds, and organic molecules. As mentioned earlier, modern-style plate tectonics might have had a indirect role in the rise of atmospheric oxygen by modifying the oxidation state of magmas and volcanic gases (Chap. 7).

The GOE marked a revolution for all the external sphere of the Earth. It promoted the development of stratospheric ozone, which provided a protective shield against solar ultraviolet radiations. Contemporaneously, methane

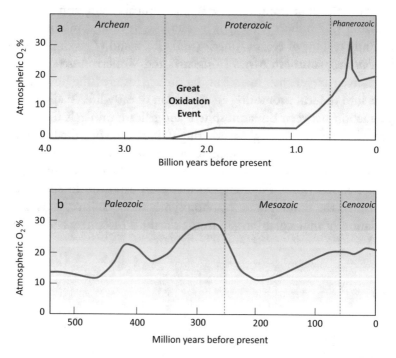

Fig. 9.1 a Variation in atmospheric oxygen from the Precambrian to present; **b** Oscillations of atmospheric oxygen during the Phanerozoic

(CH_4) and other reduced gases were oxidised, becoming minor or trace components in the atmosphere. Sulphide minerals were oxidised to sulphates, and large quantities of soluble ferrous iron in the seawater converted to ferric iron ($Fe^{2+} \rightarrow Fe^{3+}$), which precipitated on the seafloor as insoluble iron oxide and hydroxide. The accumulation of these compounds built up the widespread stratified and variably coloured rock sequences of **Banded Iron Formation** (Chap. 2) occurring in Australia, Brazil, Canada, India, Russia, South Africa, Ukraine, and the United States that comprise more than a half of the global iron reserves.

New conditions also favoured the origin and diversification of eukaryotes about 2.0 billion years ago, signalling a milestone in the evolution of life. Eukaryotes started to reproduce sexually approximately 1.2 billion years ago. Further evolution brought about the appearance of the first documented complex life forms during the Neoproterozoic, consisting of primordial plants and some soft body organisms similar to the present-day jellyfish and worms (Ediacaran fauna).

On the continents, the new oxidising conditions resulted in the formation of the so-called **red beds**, red-coloured sedimentary rocks rich in oxidised iron, which are entirely absent in older sedimentary sequences. The occurrence of red beds in the sedimentary record testifies that oxygen had accumulated in the atmosphere to oxidize iron in soil and produce red-coloured 'rusty' rocks. At the same time, some minerals typical of reducing environments, such as pyrite (FeS_2) and uraninite (UO_2), disappeared from sedimentary clastic deposits.

The mobility of lithospheric plates, collisions, assembly and breakup of continental blocks dominated the Proterozoic tectonic evolution (Fig. 9.2). The supercontinent **Columbia** (also known as **Nena**) was assembled about 1800 Ma ago and successively dismembered into several smaller continents that are presently parts of North America, Amazon, Siberia and the Baltic regions. Blocks of Columbia gathered together again at 1000 Ma to form

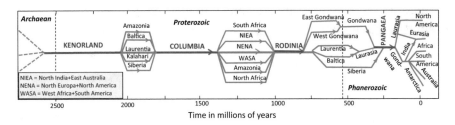

Fig. 9.2 The sequence of assembly and breakup of continents (Wilson Cycles) from Neoarchaean to present

the new supercontinent Rodinia, which in turn broke up between 775 and 750 Ma into Western Gondwana, Eastern Gondwana, Laurentia (North America and Greenland), Baltica, and Siberia; the re-aggregation of these blocks during the Phanerozoic finally formed the supercontinent Pangaea.

A sequence of fragmentation and re-assembly of supercontinents is referred to as **Wilson Cycle**. The Earth has known four of these cycles in the last three billion years[7] (Fig. 9.2). The mass of continents increases significantly during each Wilson cycle because of the addition of subduction-related magmas and sediments along the converging plate margins (Chap. 6). As a result, the last supercontinent Pangaea had an area about three times larger than Columbia.

The Proterozoic was also the aeon of the greatest impacts of cosmic bodies, still preserved in the geological record. The largest one occurred just before two billion years ago, when a meteorite, some 10–15 km in diameter, hit what is now the Vredefort area of South Africa, forming a crater about 300 km across. Another large impact occurred in the Ontario Province of Canada about 1.85 billion years ago.

The Proterozoic witnessed some of the most severe global glaciations in the Earth history. The first event occurred 2.4 billion years ago (Huronian glaciation), probably because of the GOE-related decline of atmospheric methane that reduced the greenhouse effect and lowered global temperatures. Two additional global glaciations occurred at 710 and 640 Ma during the Cryogenian (Chap. 8). These Proterozoic cooling events covered the entire Earth with an almost continuous sheet of ice, from poles to tropical or equatorial latitudes, a setting that is known as the **Snowball Earth**. However, the oceans at low latitudes remained relatively free of ice or were covered only by thin ice films. This allowed the sunlight to pass, and the photosynthetic organisms in the oceans to survive.

The latest era of the Proterozoic aeon (Neoproterozoic) experienced the appearance of Ediacaran biota, the first documented complex life forms, and a further rise in the atmospheric oxygen to values as high as 10–15% (Fig. 9.1). The high concentration of oxygen in the atmosphere might have contributed to the large-scale evolutionary phenomenon that brought about the so-called Cambrian explosion.

---

[7] Murphy et al. [8].

9.6 Phanerozoic: The explosion of Complex Organisms (541–0 Ma)

The Phanerozoic (φανερός phanerós = visible, and ζωή, zoè = life) spans the time interval between 541 Ma and the present day. It is divided into three eras, Paleozoic, Mesozoic and Cenozoic, each subdivided into various periods and stages (Table 9.1).

The Phanerozoic geodynamic evolution started with the drifting of the continental masses that resulted from the breakup of the Precambrian super-continent Rodinia. Shortly before the beginning of the Paleozoic era, about 550 million years ago, the Western Gondwana and Eastern Gondwana conti-nents merged into a single block, giving rise to the Pan-African orogeny. Successively, Laurentia and Baltica came together to form Eurasia (about 400 Ma). The collision between the latter and Gondwana and the late aggre-gation of Siberia (250 Ma) resulted in the formation of the supercontinent Pangaea, a single gigantic continental mass surrounded by the Panthalassa Ocean (Fig. 9.3a). The aggregation of Pangaea formed the Appalachian Mountains in North America and the Hercynian (Variscan) orogens in Eurasia (Chap. 3).

Pangaea began to dismember during the Middle Jurassic, about 170 Ma ago (Fig. 9.3b). This continental movement led to the opening of the Tethys Ocean, an immense sea of the Panthalassa world ocean between Eurasia and

Table 9.1 Geological time scale with indications of the main geodynamic, glacial, and biological events occurred throughout the history of the Earth. The numbers indicate ages in million years (Ma)

Aeon	Era	Period	Epoc	Ma	Main events
Phanerozoic	Cenozoic	Quaternary	Holocene	-0.0117	Homo sapiens
			Pleistocene		Glacial-Interglacial
				2.6	Genus Homo
		Neogene	Pliocene	5.3	Whales and apes
			Miocene	23	Alps
		Paleogene	Oligocene	34	
			Eocene	56	Early Himalaya
			Paleocene	66	primates
	Mesozoic	Cretaceous			Mass extinction
				145	Early flowering plants
		Jurassic		201	Breakup of Pangaea
					Mass extinction
		Triassic		252	First mammals
					Mass extinction
	Paleozoic	Permian		299	Pangaea supercontinent
		Carboniferous			Glaciation Hercynian orogeny
				359	First reptiles
					Mass extinction
		Devonian		419	First forests
		Silurian		443	Mass extinction Glaciation
		Ordovician		485	First animals on land / First fishes
		Cambrian		541	Life explosion

Aeon	Era	Ma	Main events
Proterozoic	Neo-proterozoic	541	Ediacaran fauna Panafrican
			Cryogenian glaciations orogeny (Snowball Earth)
		1000	Rodinia Supercontinent
	Meso-proterozoic		Early multicellular life
		1600	Early eukaryota Columbia Supercontinent
	Paleo-proterozoic		Vredefort metorite impact (South Africa)
			Great Oxidation Event Sexual reproduction
		2500	Huronian glaciation (Snowball Earth)
Archaean	Neo-archaean	2800	Kenorland Supercontinent
	Meso-archaean	3200	Starting modern plate tectonics
	Paleo-archaean		Oldest sedimentary rocks
		3600	Cyanobacteria - Stromatolites
	Eo-archaean	4000	Late Heavy Bombardment
Hadean			Early prokaryotes
			Earliest oceans
			Oldest terrestrial minerals
		~4600	Formation of the Earth crust

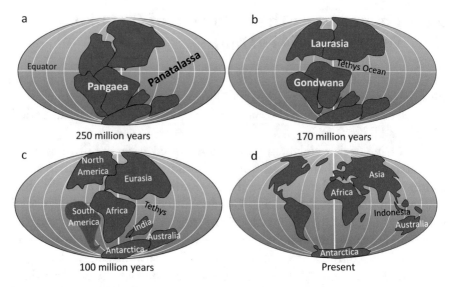

Fig. 9.3 The breakup of Pangaea from the Triassic to the present-day

Africa. Successively, the Atlantic Ocean opened by the eastern drift of the American plates (Fig. 9.3c).

During the Cretaceous, a drift inversion caused Africa and Eurasia to converge, and the Tethys Ocean to consequently shrink. The collision between Africa–Arabia and Eurasia formed the so-called Alpine orogenic belt extending from Morocco and southeastern Spain to the Alps-Apennines, Balkans, Carpathians, Turkey, and Iran. India separated from Africa and Antarctica, migrated northward and started colliding with Asia, giving rise to the Himalayan Mountain chain about 50 million years ago. At the same time, the subduction of various Pacific lithospheric plates under the western edges of South and North America resulted in the formation of the Andes and the Rocky Mountains (Laramide orogeny).

Today, Eurasia and Africa have almost joined together. The Americas continue their westward drift, while Australia is colliding with the Indonesian island arc (Fig. 9.3d). Continuing convergence is expected to lead to the aggregation of Eurasian, Africa, and Australia to form a large new continent named **Afroeuraustralasia**. In about 150 million years, the Americas will collide with Afroeuraustralasia to create a new supercontinent centred over the Arctic (**Amasia** or **Pangaea Proxima**, also pessimistically dubbed **Pangaea Ultima**), thus closing another Wilson Cycle.

Climatic changes during Phanerozoic have been a remarkable consequence of various factors. The spread of vegetation on the continents during the Ordovician and the massive burial of remains of tropical forests in the swamps of *Gondwana* and *Laurasia* provided further rise in the atmospheric oxygen to 20–30% and long-term sequestering of CO_2 in coal deposits during the Carboniferous. Changes in the atmospheric composition were the most likely cause of a long-lived glaciation during the Carboniferous-Permian. Other glaciations occurred during the Late Cenozoic, with the formation of a polar cap in Antarctica and then in the Arctic, probably an effect of the modified position of continents and the rise of the Himalayan mountain range.

9.6.1 The Spread of Animals and Plants

The extraordinary event that distinguishes the Phanerozoic from any other aeon is the flourishing of multicellular animals and plants. This started at the beginning of the Paleozoic era with the so-called **Cambrian explosion of life**, when a large number of phyla - still existing today- diversified and spread across the planet. The onset of such a revolution is marked by the first appearance in the fossil record of organisms whose bodies were protected by hard parts, such as calcite shells. The reason for this rapid and widespread proliferation of life is one of the many unsolved problems of Earth Sciences. Upsurge in atmospheric oxygen, the strengthening of the Earth's magnetic field that provided greater protection from the solar wind, genetic mutations of eukaryotic cells, and increased nutrients in the sea are some of the hypotheses put forward to explain the Cambrian explosion.

The history of life in the Phanerozoic aeon has been the subject of many excellent books and papers, so only a brief summary is given here.

The Paleozoic era (541–252 Ma) witnessed the origin, diversification, and extinction of numerous animals and plants. During the Cambrian and Ordovician periods, arthropods (**trilobites**), cephalopods, coelenterates (**corals**), and other phyla appeared and diversified. During the Ordovician, fish first appeared and then life moved from the sea onto the land. In the Silurian, the animals, plants, and mosses spread over the continents. At the end of the Ordovician, a great mass extinction affected many families of bryozoans, brachiopods, various groups of trilobites, and some marine organisms of uncertain taxonomic position called **graptolites**. Another extinction occurred at the end of the Devonian, with the almost complete disappearance of trilobites and several species of corals, conodonts, cephalopods and foraminifera. In the Carboniferous, the flourishing of forests extracted large amounts of carbon dioxide from the atmosphere; the burial of trees in immense intracontinental marshes areas resulted in the formation of large coal deposits that have sequestered huge amounts of carbon for hundreds of millions of years. The first amphibians and reptiles differentiated, and the ancestors of mammals appeared. The Paleozoic era ended 252 Ma ago, when the greatest documented mass extinction resulted in the disappearance of over 90% of the living species. This event was probably related to environmental changes induced by a large igneous province in Siberia (Chap. 3) that discharged several million cubic kilometres of magma and introduced a large poisonous gas load into the atmosphere over a few hundred thousand years.

The Mesozoic (252–66 Ma) was a relatively hot era that favoured the rapid development of great reptiles and various other species such as **ammonites** (Lower Devonian-Upper Cretaceous), nektonic marine cephalopods genetically linked to the present-day *nautilus*, the squid-like **belemnites** (Upper Triassic-Cretaceous), the sessile reef-building molluscs **rudistes** (Upper Jurassic-Cretaceous), and various types of microscopic plankton. The first **mammals** made their appearance during this time as tiny animals relegated to small ecological niches. Conifers appeared in the Triassic period; **flowering plants** made their debut in the Jurassic-Cretaceous, followed by the development of **insects**.

Two mass extinctions occurred during the Mesozoic. The first one occurred about 200 Ma ago at the transition between the Triassic and Jurassic, contemporaneously with the formation of the large igneous province of the Central Atlantic (Chap. 3); several groups of ammonites, brachiopods, amphibians, and reptiles went extinct, leaving ecological space for the explosion of the dinosaurs. The second and more famous extinction occurred 66 Ma ago at the end of the Cretaceous with the loss of many species, including the great reptiles whose disappearance opened the way to the era of mammals. The Cretaceous extinction is generally attributed to a large meteorite fall in the Chicxulub area on the northern shores of the Yucatán Peninsula, Mexico. However, recent data suggest that the decline of many animal species had started slightly earlier than the meteoric impact, possibly because of large volcanic eruptions in the Deccan province of India.

The Cenozoic (65–0 Ma) saw the development and diversifications of mammals, fishes, birds, flowering plants, and insects. Primates appeared in the Paleocene and flourished from the Eocene onward. At the transition between the Cretaceous and Paleocene, herbaceous plants emerged, as documented by the discovery of the first pollen in rocks from Africa and South America. High temperature during the Paleocene–Eocene favoured a worldwide distribution of tropical-type fauna and flora. Successive cooling brought about the thinning of forests and the expansion of savannahs and related faunas.

9.7   Finally, The Anthropocene

And then arrived the man. The genus *Homo* emerged about two or three million years before present, but the species *Homo sapiens* appeared only some three hundred thousand years ago. It is efficaciously stated that if Earth's history could be viewed as 24 h, humans would appear in the last four seconds before midnight. The blink of the eye of Earth history! Such a crucial event (at least for our species) is generally considered to have occurred in East Africa, but there are hypotheses about a simultaneous occurrence in Asia.

The presence of man and its activities have already had substantial consequences that will persist for hundreds of thousands or millions of years and, therefore, are geological in their very nature. Therefore, geologists have proposed that we are now living in the **Anthropocene,** a new epoch whose most recent stage is 'Plasticene'. As already illustrated in Chap. 8, anthropogenic activity has caused a significant and sometimes dramatic alteration of the geochemical cycles of elements that has a dramatic impact on both flora and fauna.

The chronicles of the time report that, a few centuries ago, the road between Rome and Paris ran almost entirely through forests; at present, the same route rarely enters one. The conversion of forest to agricultural or urban lands has released enormous quantities of carbon into the atmosphere, which has been estimated to be on the order of 130 billions of metric tons during only the last century and a half. Many animals, especially larger ones, became extinct, probably as a result of hunting activity. The extinction rate was particularly high between 10,000 and 8,000 years ago, when about 4% of the entire Earth's fauna, mostly large animals, disappeared. However, even in more recent times, from the sixteenth century to the present, man has wiped out thousands of plant and animal species. Such a loss of biodiversity is non-reversible and represents one of the most severe environmental problems of the Anthropocene.

Each of these effects might be considered a minor event if taken singularly. However, the modifications in the Earth system that have occurred during the Anthropocene are enormous, and the loss of living organisms is now considered the sixth mass extinction of the Phanerozoic. Unlike previous biological crises, the present extinction is not caused by catastrophic volcanic eruptions or the impact of an extraterrestrial object, but is entirely attributable to a single species, *Homo sapiens*, whose actions have the power to alter geological processes from their natural course. This ability is unprecedented in the history of life on Earth.

It is often said that anthropogenic alterations endanger the planet. Nothing could be more false. The Earth has passed much more demanding tests, always recovering brilliantly, although at its own rhythm. Anthropocene modifications put at risk many living species, including humans, but whatever the future holds, the Earth will continue.

9.8   The Gaia Hypothesis

Favourable environmental conditions greatly helped the success of complex organisms during the Phanerozoic. Therefore, the problem that needs to be addressed is whether the biosphere itself plays a decisive role in creating these circumstances, or the cause is to be found elsewhere.

There is a general agreement about the considerable impact of the biosphere on the Earth's environment, something that naturalists have been aware of since at least the nineteenth century. The Russian scientist Vladimir Vernadsky, a father of Geochemistry and Biogeochemistry, was perhaps the first to state life is a force that can drive geological processes. Studies conducted during the last century have provided support to this hypothesis, as demonstrated by discovering the Great Oxidation Event and the biological pump.

Some scientists highly emphasise the role of the biosphere, suggesting it has been a predominant factor in constructing and preserving the external environment during geological times. In other words, living organisms themselves would play a leading role in creating optimal ecosystems for their wellness.

Particularly successful and amply acknowledged is the **Gaia hypothesis** developed by the English chemist James Lovelock and the American microbiologist Lynn Margulis in the 1970s.

The Gaia hypothesis states the biosphere works in synergy with the hydrosphere, atmosphere, and pedosphere to generate an immense, self-regulating environmental system that provides long-term stability and optimal conditions necessary for the flourishing of life. The biosphere plays a key role in this complex system; living organisms would not adapt to the environment, but rather co-evolve with it. According to Lovelock, the Gaia hypothesis would find its fundamental proof in the fact that the terrestrial climatic conditions have been only mildly variable over the last 3.5 billion years, i.e. since life has appeared.[8] These conditions would be due to the action of Gaia, a *"vast being who has the power to keep our planet comfortably fit for life"*.

---

[8] Note that the Proterozoic Snowball Earth episodes were not known when the Gaia hypothesis was first formulated.

The Gaia hypothesis, some versions of which are a bit extremist, met with outstanding success while, at the same time, attracting criticism. In particular, it has been argued that the equilibrium between the external physical environment and biosphere is not by itself evidence of synergy between the two systems, but may simply mean that organisms have adapted to a habitat that has been created by other factors. To imagine the opposite, say the opponents, would be like following Professor Pangloss in thinking that the protuberance of our nose was meant for the precise purpose of leaning our glasses on it.

From a geological point of view, the dispute provides stimulating nourishment for further in-depth analysis. If the Gaia hypothesis is used as a synonym of 'Earth System', it can be integrally acceptable. However, if the idea is that the biosphere is the most elevated sphere of our home planet, Gaia becomes a typical anthropocentric conception that is at odds with the geological history of our planet.

Viewed from a broader geological perspective, the biosphere is only one wheel of a very complex mechanism that makes the Earth system work. The study of geochemical cycles teaches us that the chemical–physical state of the external environment results from the interplay of all the Earth's spheres. One of the fundamental problems of the Gaia hypothesis, at least of some extremist versions, is that the environmental impact of plate tectonics and its role in regulating the composition of the environment are overlooked if not ignored altogether. In previous chapters, it was demonstrated that extreme sodium accumulation in the water, an increase in CO_2 in the atmosphere, and water surplus at the surface are hindered by plate tectonics. And the biosphere has a little role in it!

Consider, to close the subject, the comparison between the Earth and its sister planet Venus. The Earth and Venus probably had the same atmosphere until 2.5–3.0 billion years ago, and the planets might have hosted the same types of primordial organisms. However, while the Earth evolved to the 'blue sphere' we all know, Venus became a perennial hot greenhouse planet. The contrasting evolution is probably related to the absence of water on the Venusian surface, which has not allowed CO_2 to dissolve and react with silicate minerals. Therefore, the CO_2 emitted by volcanism has accumulated in the atmosphere, leading to a 'runaway greenhouse' effect that is responsible for the extremely high temperature on the surface of Venus. On Earth, plate tectonics provides recycling of water, CO_2, and rocks continuously renovating the key ingredients for the carbon-silicate cycle. This mechanism has been keeping the terrestrial environment in a relatively stable steady-state over geological time.

In conclusion, plate tectonics, rather than life itself, seems to be the primary driver of life on Earth.

9.9   Summary

Earth's history can be viewed as a series of processes and events that transformed a glowing molten sphere into a habitable planet. This astonishing outcome was the effect of the complex interplay of both normal chemical–physical processes, such as heat loss, gravity and geomagnetism, and the occurrence of anomalous, sometimes catastrophic events such as gigantic volcanic eruptions and the fall of large meteorites.

The primordial Earth was characterised by very high internal and external temperatures related to the heat released by the decay of radioactive isotopes and meteorite impacts. Its heat slowly dissipated to outer space, and the planet solidified and differentiated to a stratified crust, mantle, and core. Solidification released gaseous substances that migrated to the external portions of the Earth and generated a primitive atmosphere rich in nitrogen, water vapour, methane and hydrogen sulphide.

When the external temperature dropped, water condensed and fell as rain, forming an immense ocean that, at some stage, completely enveloped the Earth. Water accumulation at the surface started around 4.4–4.2 billion years ago, not long after the latest stages of planet aggregation.

The slow but continuous dispersion of heat from inside the Earth led to the cooling and stiffening of the crust that broke up into various rigid blocks and gave birth to an early-stage plate tectonic regime between the Hadean and the Archaean. Modern-style plate tectonics was an outcome of further cooling of the Earth and started around 3.0 billion years ago.[9]

Early embryos of continents appeared during the Hadean. These slowly increased in size and began moving across the surface of the Earth, gathering together and then separating several times until the last supercontinent *Pangaea* was formed about 250 Ma ago. The breakup and disaggregation of Pangaea about 170 Ma ago gave rise to various continental blocks that gradually drifted into their present-day position.

Some 3.7 Ga ago, the first unicellular prokaryotic organisms appeared in aquatic environments, probably near submarine hydrothermal vents or in swampy areas on the continents. Successively, cyanobacteria started to

---

[9] Tusch et al. [9].

proliferate in shallow marine waters and, being capable of performing photosynthesis, vast quantities of oxygen slowly accumulated in the atmosphere.

About 2100 to 1600 million years ago, or perhaps earlier, unicellular eukaryotic cells appeared and successively aggregated into multicellular organisms that used oxygen for metabolism. An explosion of life occurred in the Cambrian, both quantitatively and qualitatively. Most Cambrian phyla have survived until the present time, after several mass extinctions that removed a countless number of species, but also favoured the evolution and spreading of the surviving organisms. The arrival of *Homo sapiens* some 0.3 million years ago had a profound impact on terrestrial life, affecting many animal species, dramatically modifying the external environment, and ultimately jeopardising its own possibility for survival.

9.10 Box 9.1—Geochronology: How Ages of Rocks, Fossils and Geologic Events Are Determined

In this and previous chapters, continuous reference has been made to the ages of rocks and geological events, expressing them both with numerical data (generally in millions or billions of years) and as names assigned by geologists to different time periods, such as Silurian, Triassic, Eocene, and others. Interest may therefore arise about the meaning of these terms and how ages are determined. This is a very broad topic, but is described here with considerable simplification.

The dating of any object or event can be **relative** or **absolute**. In the first case, only a time sequence is stated, i.e. what happened earlier and what happened later. Instead, absolute dating means quantifying the exact age - in number of years or else - of an event or an object. In simple terms, one can assign relative ages in a family by saying that the mother is older than her children; instead, establishing absolute ages means knowing what year each family member was born.

In geology, both relative and absolute dating is widely used. The former has been employed since the end of the eighteenth century, whereas the latter was established during the early decades of the twentieth century with the discovery of radioactivity. Before the twentieth century, the absolute age of geological objects and processes was a matter of both insufficiently constrained scientific speculation and religious interpretation of holy texts (Box 7.2).

9.10.1   Relative Age of Rocks and Fossils

Many rock series, mainly sedimentary and volcanic in origin, have a stratified structure, i.e. they are made up of superimposed layers, or strata, separated by clearly recognisable sub-parallel surfaces. Each layer was formed by a distinct geological event: a layer of sedimentary rock represents an episode of sediment deposition, a layer of volcanic ash testifies to an explosive eruption, and so on.

The succession of rock strata indicates relative ages, i.e. the time progression of their formation. The **principle of superposition of rock strata**, first formulated by the Danish naturalist Nicolaus Steno (1638–1686), states that each layer is younger than the one upon which it is resting and is older than the overlying one. Therefore, rocks lying at the bottom of a sequence are the oldest, while those resting at the top were formed later. Moreover, two layers that pass laterally one into the other are coeval. The branch of Geology that studies rock sequences is known as **Stratigraphy**.

A hypothetical rock series is schematically shown in Fig. 9.4. It comprises alternating sedimentary, lava, and volcanic ash layers. This sequence reveals that the area where this hypothetical rock series formed has experienced

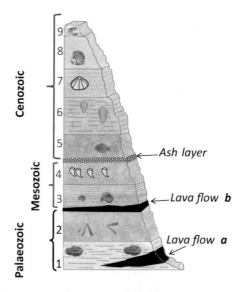

Fig. 9.4 Hypothetical series of interlayered sedimentary and volcanic rocks. Radiometric dating is only possible for volcanic rocks. The absolute ages determined for the lavas allow an absolute age to also be assigned to the sedimentary layers and their enclosed fossils

various consecutive episodes of sedimentation and volcanism in a more or less remote past.

Based on the principle of superimposition, it can be stated that ages decrease from the base (Layer 1) to the top layers of the series (Layer 9); however, nothing can be said regarding exactly when the rocks were deposited or what the time interval is between one episode of rock formation and the next. In other words, stratigraphy allows the determination of the relative ages, but not the absolute ages of rocks.

Fossils are contained in sedimentary rocks. Fossils are remains of animals that inhabited the sedimentary environments and were incorporated into the strata during deposition. In short, fossils are coeval with the host sediments. In the hypothetical series of Fig. 9.4, the lowest (and older) Layer 1 contains fossil remains of animals called trilobites. When going upward (i.e. to younger rocks), the fossil record changes gradually, with the appearance of forms looking like thin dentated sticks (graptolites), followed by a convoluted cephalopod (ammonite) and then by bivalves in the shape of cornucopia (rudist). Higher up, various fishes, gastropod and lamellibranch species appear.

It is obvious that the ages of the fossils in the layers of Fig. 9.4 decrease from the bottom to the top of the sequence, just as for the strata in which they are embedded. Therefore, trilobites are older than graptolites, which in turn are older than ammonites, and so on. Once the same fossil succession is observed with the same order in other stratigraphic series around the world, it can be concluded that the relative ages of fossils are the same worldwide.

For practical reasons, it is convenient to divide the relative ages of fossils into three successive time intervals and call 'Palaeozoic' (meaning ancient life) the era of trilobites and graptolites when Layers 1 and 2 were deposited, 'Mesozoic' (middle life) the era of rudists and ammonites (Layers 3 and 4), and Cenozoic (new life) the higher and therefore younger era during which fossils of Layers 5 to 9 were deposited. Detailed studies allow further subdivisions of these three great chronostratigraphic units to be defined, identifying periods (e.g. Cambrian, Silurian, Carboniferous, Triassic, Jurassic, etc.) and epochs (e.g. Oligocene, Miocene, etc.). At the conclusion of this process, a relative geological time scale is made (see Table 9.1), with a proper name given to each era, period, and epoch, and a one-to-one relationship established between various geochronological units and their typical fossil association.

These subdivisions have a relative chronological meaning and only tell us the rocks formed in what we call Palaeozoic are older than rocks formed in the Mesozoic and the Cenozoic. In order to have the timing in years, the absolute ages of rocks must be determined.

9.10.2 Absolute Age

An absolute age determination tells us how many years ago a mineral or rock was formed. In igneous rocks, the age of the minerals is equal to that of the host rocks; the same is often true for metamorphic rocks. Therefore, the dating of one single mineral is sufficient to obtain the age of the rock. The same is not valid for clastic sedimentary rocks because they consist of various mineral grains that come from a variety of sources whose ages may show wide differences. In this case, dating a mineral allows establishing the age of that particular grain and not the age of the rocks in which it is contained. Therefore, the absolute ages of sedimentary rocks can only be established indirectly, by dating volcanic rocks eventually interlayered with sediments.

Absolute dating techniques are mostly based on the analysis of radioactive isotopes of elements in minerals and rocks (**radiometric dating**). Isotope analyses are performed by a technique known as mass spectrometry.

Radioactive isotopes have the property of transforming by **radioactive decay** into what are called 'radiogenic' or 'daughter' isotopes. The decay rate of isotopes is determined experimentally in the laboratory; it is constant for a given isotope, but is extremely variable, by orders of magnitude, from one radioactive isotope to another. For example, the radon isotope ^{218}Rd (or radon-218) rapidly decays to the polonium isotope ^{214}Po, reaching one half of its initial abundance (**half-life**) after 38 ms; another radium isotope ^{222}Rd decays to polonium isotope ^{218}Po with the half-life of a few days; the radioactive isotope of carbon or radiocarbon (^{14}C) has a half-life of 5370 years and decays into a stable nitrogen isotope (^{14}N); isotopes of K, Rb, U and Th (^{40}K, ^{87}Rb, ^{238}U, ^{232}Th) decay very slowly to various isotopes, showing half-lives of billions of years.

The basic principles of radiometric dating are simple, although the practical difficulties can be severe. Imagine having an aquarium hosting various types of fishes. One of them, say the goldfish, converts into sunfish at a constant rate, for example, one transformation per day. Suppose the aquarium is sealed so that no goldfish or sunfish can enter or exit. In that case, the number of days that have passed since the closure of the aquarium can be easily determined at any time by counting the number of either goldfish or sunfish.

Radiometric dating obeys the same principle. It can be carried out on minerals or rocks, provided they contain radioactive isotopes. Measuring the abundance of radioactive parent isotopes (corresponding to the goldfish) and radiogenic daughter isotopes (corresponding to the sunfish) allows us to calculate, from their decay rate, how many years ago the parent isotopes were locked inside the crystal lattice of the mineral, i.e. when the mineral crystallised.

The radioactive isotopes used for dating minerals include uranium, thorium, potassium, rubidium, among others. The choice of the isotope to be used depends on various factors, which will not be considered in this brief account.

The absolute dating of minerals and rocks allows the absolute age for fossils and geological periods and eras to be indirectly determined. To explain this point, consider Fig. 9.4 again. The stratigraphic sequence contains fossiliferous sedimentary layers, lava flows, and a layer of volcanic ash; however, absolute ages can be established only for the volcanic rocks. Radiometric dating of '*lava flow a*' allows an absolute age for the entire Layer 1 and the enclosed fossils to be established, since there is a continuous lateral transition between lava and fossil-bearing sediments, which means they all are coeval. The radiometric dating of '*lava flow b*' and '*ash layer*' allows the absolute ages of the transitions from the Palaeozoic to Mesozoic eras, and from the latter to the Cenozoic, to be established.

Similar studies of many rock series from various parts of the world have led to the determination of the absolute ages being assigned to almost all the fossils and the different periods and eras, thus appending a number of years to chronostratigraphic units and their fossil associations.

References

1. Allégre CJ, Manhés G, Göpel C (1996) The age of the Earth. Geochim Cosmochim Acta 59:1445–1456
2. Mojzsis SJ, Harrison TM, Pidgeon RT (2020) Oxygen-isotope evidence from ancient zircons for liquid water at the Earth's surface 4,300 Myr ago. Nature 409:178–181
3. Nutman AP, Bennett VC, Friend CRL, Van Kranendonk MJ, Chivas AR (2016) Rapid emergence of life shown by discovery of 3700-million-year-old microbial structures. Nature 537:535–538
4. Dodd MS, Papineau D, Greene T, Slack JF, Rittner M, Pirajno F, O'Neil J, Little CTS (2017) Evidence for early life in Earth's oldest hydrothermal vent precipitates. Nature 543:60–64
5. Condie KC, Pease V (eds) (2008) When did plate tectonics begin on planet Earth? Geol Soc Am Spec Paper 440, 295 p
6. Taylor SR (1967) The origin and growth of continents. Tectonophysics 4:17–34
7. Dhuime B, Wuestefeld A, Hawkesworth CJ (2015) Emergence of modern continental crust about 3 billion years ago. Nat Geosci 8:552–555

8. Murphy JB, Keppie JD, Haynes A (eds) (2009) Ancient Orogens and Modern Analogues. Geol Soc Spec Publ 327, 488 p
9. Tusch J, Münker C, Hasenstab E, Jansen M, Marien CS, et al (2021) Convective isolation of Hadean mantle reservoirs through Archean time. Proc Natl Acad Sci 118 (2)

10

Epilogue

O malheureux mortels! ô terre déplorable!
O de tous les mortels assemblage effroyable!
D'inutiles douleurs éternel entretien!
Philosophes trompés qui criez: "Tout est bien"
Accourez, contemplez ces ruines affreuses
Ces débris, ces lambeaux, ces cendres malheureuses...
François Marie Arouet de Voltaire, Poème sur le désastre de Lisbonne (1756)

10.1 The Earth Anomaly

The Earth is an extraordinary planet, the only one in the universe so far known possessing the qualities that make it suitable for complex life. The astrophysical and geological conditions that led to this outcome are numerous and often interdependent.

The key condition to sustain complex life is the presence of surface liquid water and rather uniform climatic conditions. Both outcomes are the effect of several astronomical factors, such as the distance from the Sun, the shape of the orbit, the axial inclination, and the mass of the planet.

The distance from the Sun is appropriate for the Earth to have liquid water on its surface. Mercury and Venus, the planets that are closer to the Sun, are exceedingly hot and surface water evaporates; planets far distant from the Sun are exceedingly cold so that water freezes.

The near-circular orbit of the Earth allows for a relatively constant inflow of heat between perihelion and aphelion, the points where the Earth's orbit is at the closest and the farthest distance from the Sun. If the orbit were more elliptical, the intensity of incident solar radiation would be highly variable throughout the year, causing continuous alternation between freezing and excessively hot periods. Such extreme variation in temperatures would undoubtedly prevent water condensation and hinder the development of complex life.

The 23.4° inclination of the Earth's axis is another important factor affecting climatic stability. This axial tilt is why winter and summer occur simultaneously, opposite to each other in the northern and southern hemispheres. This alternation of the seasons helps solar energy dissipate rather uniformly over much of the planet, thereby levelling global temperatures.

The rotation of the Earth around its axis is fast, providing for a rapid alternation of day and night. Such short days and nights are essential to an even distribution of solar energy on the planet.

The presence of the Moon acts as a stabilising force on the Earth's axis of rotation. Without the Moon, this axis would be subject to excessive wobbling, creating erratic climatic conditions. As explained in Chap. 9, the origin of the Moon and the Earth's speed of rotation are the ultimate effects of the giant collision of early Earth and the proto-planet Theia. Some argue that such a titanic collision also provided the energy to the molten metallic core to start convection and give rise to the geomagnetic field: a very 'fortunate' cataclysm indeed.

Due to their powerful gravitational forces, giant planets, especially Jupiter, attract many cosmic bodies. Without Jupiter, the Earth would be frequently struck by meteorites and comets, with consequent catastrophic effects.

Finally, the Earth has the right mass to retain water and oxygen. A smaller mass would have allowed these chemical components to disperse into space. Likewise, a greater mass would have kept exceedingly large amounts of hydrogen that would combine with oxygen to produce water, thus drastically reducing the amount of free oxygen in the Earth's atmosphere. In the absence or scarcity of oxygen, the atmosphere would not have an ozone layer to protect the surface from ultraviolet radiation coming from the Sun.

Many geological factors also contribute to making the Earth a habitable planet. Its stratified structure, formed under the action of gravity and magmatism, represents a crucial geological feature and is the primary condition required for the most important processes, especially plate tectonics.

The presence of a metallic core contributes to the specific density of the planet; the liquid outer core is the source of the geomagnetic field that provides essential protection against lethal cosmic radiation; a mechanically plastic mantle facilitates the transport of heat and matter towards the lithosphere prompting cooling and differentiation of the planet; the external fluid portions of the Earth constitute the environments that host life.

Plate tectonics maintains the steady-state condition of the Earth. It reintroduces into the mantle many chemical components, which would otherwise abnormally accumulate on the surface. The water transported into the mantle contributes to its unusual mechanical characteristics, allowing large-scale convection and making plate mobility possible.

Subducted water also increases the oxidation state of the upper mantle. Consequently, gases emitted by volcanoes contain relatively low quantities of reducing chemical substances, which lowers the consumption of molecular oxygen and favours its accumulation in the atmosphere.

Continents are a fundamental outcome of plate tectonics. They form almost all of the emergent land that hosts a diversity of higher forms of life. Moreover, continents are the primary source of nutrients for the marine realm; without the enormous masses of rocks exposed to weathering on the continents, the oceans would be deficient in nutrients and would likely be an immense liquid mass containing little or no living organisms.

Unfortunately, plate dynamics cause seismicity and volcanism, two of the most dreadful geological phenomena that cause destruction and death on a global scale, as seen during the emplacement of the Large Igneous Provinces. Magmatism, however, has played a vital role in constructing a stratified planet and building up the continents and atmosphere. Seismicity is an inevitable consequence of plate mobility. Seismicity and volcanism are, metaphorically, the price to pay for a habitable planet. Humanity has to coexist with them, as it potentially has the intellectual means to minimise risks, even if it does not always manage to do so effectively.

In conclusion, many factors have contributed to creating a planet suitable for complex life, in particular, for the human species. Some people contend that such an outcome results from the natural action of geological processes and many accidental catastrophic events; others have a more finalistic vision and believe that a superior intelligence has organised everything with the precise purpose of hosting, at the end of a long history, human beings. However, the latter is a suggestive anthropocentric idea that fails to meet the verifiability and falsifiability criteria. Therefore, following the philosopher of science Karl Popper (1902–1994), this belief cannot be considered a scientific hypothesis.

10.2   The Best of All Possible Worlds

The coexistence of contrasting phenomena in nature has been a topic of philosophical speculation by some of humanity's most brilliant minds. At the beginning of the eighteenth century, the German philosopher and mathematician Gottfried Wilhelm Leibniz published *Essais de Théodicée sur la bonté de Dieu, la liberté de l'homme et l'origine du mal* (Essays of theodicy on God's goodness, man's freedom and the origin of evil). In this treatise, Leibniz discussed the badness of the world and the goodness of God, that is, the theodicy, one of the most arduous subjects of philosophy and theology. The central question of the theodicy is as follows: "If God exists and possesses supreme goodness and wisdom, where does evil come from?". In Latin, it is simply put forth as *Si Deus unde malum?* Leibniz concluded that our world was created in such a way to be the best possible place for man to live. Such an optimistic proposition gathered some consensus, but also met fierce opposition.

The strongest dissent was expressed by François-Marie Arouet—better known as Voltaire—who ridiculed Leibniz's point of view in his *Candide, ou l'Optimisme* (Candid or Optimism), in which he tells of the pertinacious optimist Professor Pangloss, who obstinately goes on with considering our world as the best possible one, despite the terrible chain of misadventures striking himself and his companions. Voltaire's opposition became more implacable after the catastrophic earthquake that struck Lisbon on 1st November 1755 and wreaked havoc across the Portuguese capital. How is it possible, Voltaire wondered, that disasters of such a magnitude could happen in the best of all worlds?

The question makes sense only if one believes that man is the culmination of evolution (or creation or intelligent design, or else) and that everything must be aimed at his wellness. However, such an overarching philosophy has lost its foundation since Charles Darwin has radically changed our view of the world and inflicted a lethal blow to the conception of man's centrality.

The present, scientifically grounded opinion of many scholars is that the Earth is only one of the many planets dispersed in the known universe, which experienced a unique combination of physical–chemical processes and random events that made it fit for life. In short, the Earth is simply a very 'lucky' planet rather than the product of an intelligent design aimed at providing man's goodness. Nothing has been created for a particular scope, and the human species and other organisms have adapted to Earth's different environments to survive. Perhaps, it could make some sense to say that the organisms surviving on Earth are the best possible ones because those which were inadequate lost their evolutionary struggle. So far, humans have adapted very well and have spread to every corner of the planet.

Such a success will last as long as humankind remains in substantial equilibrium with the Earth's changing environment. Should this balance be upset, even we humans could disappear, as it has happened to many other animal species in the remote and near past of the Earth's history.

The causes that can threaten life-supporting conditions or the survival of the human species are linked not only to natural events, but also to man's action itself. We can do little to change the course of natural phenomena, but we can do a lot to improve our behaviour and preserve the optimal conditions for our environment.

10.3 Historical Times, Geological Times

The appearance and subsequent proliferation of humankind are among the most significant events in the recent history of the Earth. *Homo sapiens* arose 300,000 years ago in Africa and, from there, spread across the planet, reaching the current staggering number of 7.8 billion people. The expansion of animal species is a sign of adaptability to the surrounding environment. However, extreme success can produce environmental changes that can jeopardise the conditions for life and even put the existence of a species in peril. This principle also applies to *Homo sapiens*.

It has happened several times during the Earth's history that living organisms have modified their habitat. It occurred during the Proterozoic when the activity of photosynthetic algae led to an increase in the atmospheric oxygen levels. Other examples include the repeated oscillations of atmospheric CO_2 and oxygen as a result of the decline or development of vegetation on land and in the sea. These processes took place very slowly, coming to completion in hundreds of thousands or millions of years. The human time scale is much faster, and even the most long-lived empires last a maximum of centuries. Anthropogenically-driven changes of the external environment are anomalous because, although affecting geological systems, they take place over very short periods, in the order of tens or a few hundreds of years. In other words, geological processes that need millions of years to complete are highly accelerated by human activity and are accomplished over human or historical time scales. The contrast between historical and geological timing of processes is the crucial aspect of the present-day environmental changes.

10.4 Technology, Natural Resources and Environmental Impact

Modern society needs increasing amounts of energy and raw material to preserve and improve its lifestyle. At present, most of the energy comes from the burning of fossil fuels extracted from crustal sediments, as discussed in Chap. 8. Almost all of the raw materials required by our technology come from the Earth's crust as well. The list includes the most common metals such as iron and copper, and rare but precious elements such as lithium, beryllium, chrome, cobalt, gold, platinum, the rare earth elements, tungsten, osmium, tantalum, niobium, lead, and many others. These elements are indispensable for building bridges, railways, electrical systems, television sets, electronics and computers, mobile phones, batteries, and countless other everyday objects. Without them, modern technological civilisation would not be possible.

In general, metals are present in the Earth's crust in such low concentrations that they can only be extracted at very high economic and environmental costs. However, there are certain restricted zones in the crust where elements of economic importance are concentrated in high quantities. These mineral deposits are produced as a result of magmatism, metamorphism, or sedimentation.

As discussed in Chap. 8, the exploitation of mineral deposits by humans is a perturbation of the natural geochemical cycles, which enormously accelerates the element transfer from the lithosphere to the external environment. Hydrocarbons, coal, and metals would remain in the crust for hundreds of thousands or millions of years before entering the biosphere–atmosphere–hydrosphere system. However, societal need has accelerated this transition, accomplishing it in a few decades or centuries. Even worse, exploitation of natural resources is sometimes performed without environmental sensitivity or unethically, by inflicting incalculable damages upon the environment and cynically exploiting child and forced labour.

Disturbance in the natural cycle of the elements has alarming consequences. One of the negative effects is the drastic impoverishment of ore reserves, many of which are declining or nearing exhaustion. A second outcome is the introduction of anomalous quantities of foreign and often harmful substances into water, air, soil, and biota. Besides carbon and its compounds, particularly plastics, environmental pollution involves many potentially toxic elements such as phosphorus, lead, arsenic, mercury and radioactive elements. The use of natural resources is inevitable if current levels of technology are to be maintained or enhanced. However, it is necessary to be aware that the wild extraction and abuse of these materials result in disruption of the geological processes, resulting in harmful anomalies in the environment that will take geological times to recover.

10.5 *Natura, Non Nisi Parendo Vincitur*

The history of climate summarised in Chap. 8 shows that the Earth's environment has been changing continuously. During the Holocene, temperature oscillations have been minimal and episodes of relatively hot climate alternated with cool periods. According to most popular models, a hot climate similar to that at present occurred in prehistoric times, around the beginning of the Christian era and during the Middle Ages, without producing dramatic consequences. Such a statement could lead to the conclusion that the present-day alteration of the environment is a typical flickering of a dynamic planet, and that there is nothing new under the sun.

This observation overlooks three important aspects of the present time. One is the rate at which current environmental change is taking place, which has little or no precedents in the geological record. A second important aspect is that the Earth has never hosted a human population of such magnitude. Finally, the particular organisation of modern society, characterised by maximum complexity but also by maximum vulnerability, has to be taken into the uttermost consideration. Some geological or even cosmic events (e.g. the solar flares) that did not have any significant effect in the past, may now prove dramatic in the hypertrophic and complex society we have built. An example is the eruption of the Icelandic volcano Eyjafjallajökull. It was a modest phreatomagmatic eruption of little volcanological significance, which in other times would have gone completely unnoticed. Yet, in the spring of 2010, the ash cloud blew over Europe and the North Atlantic, stopping the air services for several days and resulting in very high economic and social costs.

In essence, the problems related to ecological factors are strongly enhanced by the enormous growth of the world population and the extreme complexity of human society, which turns any modification in the environment, even the modest one, into events with devastating consequences. The question is, therefore, about what we should do to maintain the Earth's environment in the best possible condition and how all that we have learned about geological phenomena can help in making the correct choices.

In this author's opinion, such an issue can only have one answer: give utmost respect to the course of geological processes and, in particular, the geochemical cycles of the elements; avoid traumatic actions; and adopt measures to rectify anomalies whenever they occur. These objectives are achieved by disposing of waste in such a way as to be compatible with the natural cycle of the elements to the maximum extent possible. We must minimise energy consumption from non-renewable sources, use the abundant energy from wind, geothermal heat and solar radiation, return chemical elements extracted from natural deposits back to their starting places in the lithosphere, follow respectful practises in agriculture, transport and industry, recycle what is no longer used instead of disposing it in the environment, and protect every sector of the land, the sea and the biosphere.

Following these rules does not mean giving up our present situation and style of life and return to 'the good old times'. Just the opposite! It does mean that we have to take equal care to respect the nature that we have given to exploiting it. Humankind has to make an intelligent, efficient, targeted use of the scant resources Earth has accumulated during its long history. We have to spend some of our energy and resources to leave order in the nature we are exploiting, instead of focusing on the unconstrained abuse of the Earth's richness and resources. This should not be done only for moral reasons, but because this is our own interest and that of our children and grandchildren.

Perhaps the most valuable suggestion regarding what the human attitude should be towards the Earth's system comes from a famous sentence by the English philosopher, scientist and statesman, Sir Francis Bacon (1561–1626): *Natura, non nisi parendo vincitur* (Nature, is only subdued by submission). This aphorism, which should be the motto of geologists and naturalists, contains two fundamental precepts. The first one is that human action may be in opposition to nature; the second one is that under no circumstances should nature be harassed, but must be respected and eventually modified only by observing its laws.

Respect for nature presupposes its knowledge not only by specialists, but particularly by the public in general. This is the area to which this book intends to give a modest contribution.

Correction to: Air, Water, Earth, Fire

Correction to:
A. Peccerillo, *Air, Water, Earth, Fire*,
https://doi.org/10.1007/978-3-030-78013-5

The original version of the book was inadvertently published without updating the following corrections, which have now been updated:

Page iv: Cover illustration: The Earth: Photograph by NASA; Yellowstone volcano: Photograph courtesy Dr. Maria Fatima Viveiros, University of Azores.

Page xi: in Heading 6.6, "Starts" has been changed to "Start".

Page 16: in line 1 from top, "ring" has been changed to "string".

Page 27: in line 4 from top "pH-8" has been change to "pH~8".

Page 37: in line 2 after the chemical reaction (2.8), "reaction (6)" has been changed to "reaction (2.6)".

Page 37: in line 6 after reaction (2.8), "reactions (7) and (9)" has been changed to "reactions (2.6) and (2.8)"

Page 120: in Heading 6.6, "Starts" has been changed to "Start".

Page 149: in line 4 from top, "In 1858" has been changed to "On July 1st 1858", and at line 7 from top, "on July 1st 1858" has been deleted.

Page 174: in Footnote n. 21, "&#xF0B4" has been changed to "x".

Page 205: in line 15 from top, the words lava and ash should have been changed to italics.

The updated version of the book can be found at
https://doi.org/10.1007/978-3-030-78013-5

Page 208: in line 3 from bottom, "geological" has been changed to "processes".

The book and the chapters have been updated with the changes.

Further Reading

1. Ball P (1999) H_2O: A biography of water. Weidenfeld & Nicolson, 387 p
2. Benton MJ (2008) When life nearly died: The greatest mass extinction of all time. Thames and Hudson, 336 p
3. Bryson B (2003) A short history of nearly everything. First Trade Paperback Edition, 544 p
4. Emiliani C (1992) Planet Earth: Cosmology, Geology, and the evolution of life and environment. Cambridge University Press, 736 p
5. Francis P, Oppenheimer C (2003) Volcanoes. Oxford University Press, 536 p
6. Frisch W, Meschede M, Blakey RC (2011) Plate Tectonics. Continental drift and mountain building. Springer, 212 p
7. Gribbin J (2012) Planet Earth. Oneworld Publications, 175 p
8. Harmon RS, Parker A. (2011) Frontiers in Geochemistry. Contribution of Geochemistry to the Study of the Earth. Wiley-Backwell, 263 p
9. Hazen M. R. (2012) The Story of Earth: The first 4.5 billion years, from stardust to living planet. Viking Press, 320 p
10. Jones S (2015) Introducing sedimentology. Dunedin Academic Press, Edinburgh, 86 p
11. Kearey P, Klepeis KA., Vine FJ (2009) Global tectonics. Wiley-Blackwell, 482 p
12. Knoll HA (2004) Life on a Young Planet: The First three billion years of evolution on Earth. Princeton University Press, 288 p
13. Lana N (2004) Oxygen : The molecule that made the world. Oxford University Press, 384 p
14. Lenton T, Watson A (2014) Revolutions that made the Earth. Oxford University Press, 423 p
15. Lopes R (2010) Volcanoes. Oneworld Publications, 150 p

© The Editor(s) (if applicable) and The Author(s), under exclusive license to Springer Nature Switzerland AG 2021, corrected publication 2022
A. Peccerillo, *Air, Water, Earth, Fire*,
https://doi.org/10.1007/978-3-030-78013-5

16. Lunine JI (2013) Earth. Evolution of a habitable world. Cambridge University Press, 318 p

17. Macdougall D (2008) Nature's clocks: How scientists measure the age of almost everything. University of California Press, 288 p

18. MacLeod N (2013) The great extinctions. Natural History Museum, London, UK, 208 p

19. Marshak S (2007) Earth: Portrait of a planet. Norton & Company, 880 p

20. Merril RT (2010) Our magnetic Earth. The University of Chicago Press, 261 p

21. Molnar P (2015) Plate Tectonics: a very short introduction. Oxford University Press, 160 p

22. Park G (2015) Mountains. The origin of the Earth's mountain system. Dunedin Academic Press, 212 p

23. Park G (2016) Introducing geology. Dunedin Academic Press, 134 p

24. Rudwick MJS (2014) Earth's deep history. How it was discovered and why it matters. University of Chicago Press, 360 p

25. Saraceno P (2007) Il caso Terra. Mursia, Milan (in Italian), 376 p

26. Schlesinger WH, Bernhard ES (2013) Biogeochemistry. An analysis of global change. Elsevier, 672 p

27. Stager C (2011) Our future Earth. Duckworth Overlook, 284 p

28. Van Andel TH (1985) New views of an old planet. Continental drift and the history of the Earth. Cambridge University Press, 324 p

29. Vita VC, Fortes D (2013) Planetary geology. An introduction. Dunedin Academic Press, 206 p

30. Waltham D (2014) Lucky planet. Icon Books, 225 p

31. Ward P, Kirschvink J (2015) A new history of life. Bloomsbury, 391 p

32. Webb J (ed) (2018) This is planet Earth. New Scientist, 274 p

33. White MW (ed) (2018) Encyclopedia of geochemistry. A comprehensive reference source on the Chemistry of the Earth. Springer, 1557 p

34. Whitehouse D (2015) Journey to the centre of the Earth. Weidenfeld & Nicolson, 270 p

Index

Phanerozoic

Precambrian

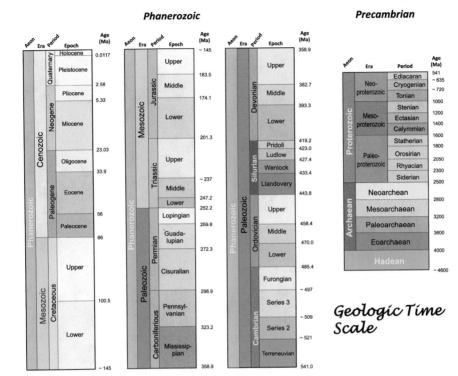

Geologic Time Scale

Printed in the United States
by Baker & Taylor Publisher Services